D1305201

Gulf of Mexico Origin, Waters, and Biota

Volume 2, Ocean and Coastal Economy

Harte Research Institute for Gulf of Mexico Studies Series
Sponsored by the Harte Research Institute for Gulf of Mexico Studies,
Texas A&M University–Corpus Christi
John W. Tunnell Jr., General Editor

Gulf of Mexico Origin, Waters, and Biota

Volume 2, Ocean and Coastal Economy

Edited by **James C. Cato**

Texas A&M University Press
College Station

This paper meets the requirements of ANSI/NISO Z39.48-1992 (Permanence of Paper).
Binding materials have been chosen for durability.
⊗

Library of Congress Cataloging-in-Publication Data

Gulf of Mexico origin, waters, and biota / [edited by John W. Tunnell, Jr.,
Darryl L. Felder, and Sylvia A. Earle] — 1st ed.
    v. cm. — (Harte Research Institute for Gulf of Mexico Studies
    series)
Includes indexes.
Taken from the Harte Research Institute for Gulf of Mexico Studies website: Gulf
of Mexico origin, waters, and biota, is an updated and enlarged version of the
Gulf of Mexico: its origin, waters, and marine life, first published by U.S. Fish and
Wildlife Service in Fishery bulletin, v. 89, 1954. Contents: V. 1. Biodiversity / edited by
Darryl L. Felder and David K. Camp ISBN-13: 978-1-60344-094-3 (cloth : alk. paper)
ISBN-10: 1-60344-094-1 (cloth : alk. paper)
1. Mexico, Gulf of. 2. Marine biology—Mexico, Gulf of. 3. Geology—Mexico, Gulf
of. 4. Oceanography—Mexico, Gulf of. I. Tunnell, John Wesley II. Felder, Darryl L.
III. Earle, Sylvia A., 1935– IV. Camp, David K. V. Series.
QH92.3.G834 2009
578.77'364—dc22
                                                                    2008025312

Vol. 2, Ocean and Coastal Economy
ISBN-13: 978-1-60344-086-8
ISBN-10: 1-60344-086-0

Contents

Foreword:
Fifty-Year Update of Bulletin 89

Just over fifty years ago, a group of prominent marine scientists of their day agreed to begin work on a digest of existing knowledge on the Gulf of Mexico. The effort was proposed by Lionel A. Walford of the U.S. Fish and Wildlife Service and Waldo L. Schmitt of the U.S. National Museum of Natural History during a meeting of the Gulf and Caribbean Fisheries Institute in Miami. Paul S. Galtsoff of the Fish and Wildlife Service agreed to coordinate the project, the magnitude of which he subsequently found far exceeded his expectations. However, three years of effort by fifty-five contributors and additional months of editing resulted in the 1954 publication of a classic reference work entitled *Gulf of Mexico—Its Origin, Waters, and Marine Life* as Fishery Bulletin 89, Fishery Bulletin of the Fish and Wildlife Service, volume 55 (Galtsoff, 1954). The table of contents for the volume appears at the end of this foreword. On the title page of the work is an explanatory note that it was "Prepared by American scientists under the sponsorship of the Fish and Wildlife Service, United States Department of the Interior" and that the effort was "Coordinated by Paul S. Galtsoff," who is generally indicated as the editor in bibliographic references to it.

For more than fifty years this reference volume—commonly referred to simply as "Bulletin 89" by hosts of marine scientists, agency personnel, and students familiar with it—has provided a benchmark on which to build. Chapters on the history of exploration, geology, meteorology, physical and chemical oceanography, biota, and pollution remain extremely valuable as reference works, some now primarily for historical context. Counted among the contributors were the most distinguished North American marine scientists of their day, and visibility for a number of them was further enhanced by the extensively cited chapters they contributed to this volume. The group included the most qualified federal agency scientists, museum curators, marine laboratory investigators, and university professors who could be assembled. It broadly represented taxonomic authorities selected to cover almost every possible biotic group, with acknowledged omission of some groups for which willing expertise could not be found.

The original Bulletin 89 was heavily slanted toward biology, reflecting the focus of that era. A page count by topic reveals 63% biology (plant and animal communities 10%, biota 53%), oceanography 11%, geology 9%, history 6%, pollution 4%, meteorology 2%, and the index 5% (see table of contents).

At the time of this writing, only one of the fifty-five original contributors remains alive. However, all the original contributors, and especially the far larger number of students they mentored, have contributed to a massive body of information on the Gulf of Mexico since 1954. In addition to this core group, a number of other workers—many now in laboratories, agencies, and university programs that did not exist fifty years ago—have made tremendous contributions to the baseline knowledge of the Gulf of Mexico since publication of the original volume.

In September 2000, Ed Harte, former owner of the *Corpus Christi Caller-Times* and Harte-Hanks Publishing, gave Texas A&M University–Corpus Christi (TAMU–CC) a $46 million endowment to establish a research institute to study and conserve the Gulf of Mexico. Soon afterward, then President Robert Furgason obtained an additional $18 million from the State of Texas for a building to house the institute. Sylvia Earle, whose book *Sea Change* (Earle, 1995) had inspired Harte's gift, was invited to chair the Advisory Council. She and Bob Furgason then began establishing a world-class Advisory Council of leaders in science, academics, conservation, government, and industry. John W. ("Wes") Tunnell Jr. was subsequently asked to serve as associate director to assist in guiding the institute development process, to coordinate construction of the new building, and to develop a new doctoral-level graduate program with other TAMU–CC faculty. This newly developing organization was given the name Harte Research Institute (HRI) for Gulf of Mexico Studies (Tunnell and Earle, 2004). Further information about HRI and TAMU–CC can be found at their respective websites (www.harteresearchinstitute.org and www.tamucc.edu).

After two HRI Advisory Council meetings, Wes Tunnell was encouraged to develop some "early" projects during the formative years of HRI in order to get a jump start on its mission of developing a cooperative and collaborative research institute focused on the long-term sustainable use and conservation of the Gulf of Mexico. From that emerged a multi-year, tri-national initiative called *The Gulf of Mexico—Past, Present, and Future* (Tunnell et al., 2004). The initiative had nine components, all of which included participation by the three countries surrounding the Gulf of Mexico: Cuba, Mexico, and the United States. Three of the nine components centered on determining current knowledge about the Gulf of Mexico ecosystem: (1) State of Knowledge Workshop, (2) Biodiversity of the Gulf of Mexico Project (Tunnell 2005), and (3) preparation of a fifty-year update of Bulletin 89. The biodiversity and Bulletin 89 projects were initially conceived and discussed by Sylvia Earle, Wes Tunnell, and Darryl Felder in late 2001 and early 2002. Concept development continued through early 2003, when a steering committee was formed to develop ideas further and establish an implementation strategy. Steering Committee members included Fernando Alvarez, Bill Bryant, Ernesto Chávez, Luis Cifuentes, Steve Dimarco, Quenton Dokken, Sylvia Earle (co-chair), Elva Escobar, Ernie Estevez, Darryl Felder, Suzanne Fredericq, María Elena Ibarra, Chuck Kennicutt, Paul Montagna, Marion Nipper, Worth Nowlin, Manuel Ortiz, David Pawson, Nancy Rabalais, Wes Tunnell (chair), and Gene Turner.

Overall objectives for the projects were as follows:

- to produce an updated Bulletin 89
- to provide a benchmark work by the leaders in the field at the beginning of the twenty-first century
- to provide a synthesis of all work to date to the scientific, management, business, and policy communities to encourage an ecosystem view of the Gulf of Mexico
- to encourage cooperation and collaboration among U.S., Mexican, and Cuban scientists working in the Gulf of Mexico, and
- to determine information gaps in knowledge of the Gulf of Mexico, so targeted research can be encouraged to fill those gaps.

The State of Knowledge Workshop, held 14–16 October 2003, became the kickoff for the biodiversity and Bulletin 89 projects. The fifty-year update of Bulletin 89 grew from one volume to seven, broadly including geology, physical and chemical oceanography, biota, anthropogenic issues, ecosystem-based management, and socioeconomics. Likewise, the effort was expanded from fifty-five authors in 1954 to more than two hundred for the new effort. Most of the knowledge gained and presented in the original Bulletin 89 was from research cruises and expeditions to the Gulf during the late nineteenth and early twentieth centuries and from a few fledgling marine science labs and oceanography programs started in the early to mid-1900s, but massive efforts have followed in their wake.

At the dawning of a new century, researchers at marine labs and universities encircle the Gulf in Cuba, Mexico, and the United States (see www.gulfbase.org), and instrumentation, technology, and communication have greatly expanded our knowledge of the Gulf. The U.S. Environmental Protection Agency's Gulf of Mexico Program has identified priority problems affecting the northern Gulf of Mexico, and the agency recently published the research needs of that region (EPA, 2002). A network of United Nations organizations has declared the Gulf of Mexico as one of 64 large marine ecosystems in the world (Kumpf et al., 1999). The U.S. Commission on Ocean Policy released its report, *An Ocean Blueprint for the 21st Century*, in August 2004 listing 212 recommendations for actions to care for and manage U.S. coasts and oceans better. President George W. Bush subsequently issued his Ocean Action Plan response in December 2004, singling out the Gulf of Mexico as a region of special concern.

Next HRI sponsored the first State of the Gulf of Mexico Summit. This three-day summit, held 26–30 March 2006, was attended by 450 invited guests from the United States and Mexico, and it focused on governance, catastrophic events, sustainability, economics, public health, and the environment (Tunnell and Dokken, 2006). Future summits are planned at regular intervals, perhaps every three years, at key locations around the Gulf of Mexico.

While the challenge was daunting, an update of Bulletin 89 was long overdue. As the fiftieth anniversary of its publication has passed, the range and scope of primary literature sources on the Gulf of Mexico have become so expansive as to be all but unmanageable for most workers. For almost all subject areas, no authoritative digests centered on the Gulf of Mexico have appeared since Bulletin 89. Yet many treatments in that work are clearly outdated and are of limited value except as historical starting points. Furthermore, it was urgent to begin compilations for this updated digest before the marine science community sustained further loss of continuity in expertise. We have already lost all but one of the original contributors to Bulletin 89, the passage of fifty years has claimed a large number of the subsequent generation of workers, and others are late in their careers. This is perhaps most evident in what has become a very limited pool of qualified systematists to draw upon for expertise concerning diversity and taxonomy of the Gulf of Mexico biota. It is noteworthy that one remarkable scientist who contributed to the original effort, the late Frederick M. (Ted) Bayer, also coauthored a chapter for one volume (*Biodiversity*) immediately prior to his passing.

We collectively thank all contributors to this immense effort, especially the editors and coordinators of each volume for their multi-year commitments to this project. Many users will benefit for decades to come because of their dedicated efforts. We especially thank Ed Harte for putting in place the infrastructure needed to undertake a project of this magnitude, and we offer these volumes as an early step in response to his charge—"make a difference."

—John W. Tunnell Jr.
Darryl L. Felder
Sylvia A. Earle
Bulletin 89: Fifty-Year Update Series Coordinators

Bulletin 89 Table of Contents (Abbreviated)

Gulf of Mexico—Its Origin, Waters, and Marine Life
Paul S. Galtsoff, Coordinator
Fishery Bulletin 89, 1954

Fishes by Luis R. Rivas (3 pp.)
Commercial fishes by George A. Rounsefell (6 pp.)
Sea turtles by F. G. Walton Smith (3 pp.)
XVIII. Birds by George H. Lowery, Jr., and Robert J. Newman (22 pp.)
XIX.   Mammals by Gordon Gunter (9 pp.)
XX.   Pollution of water (21 pp.)

References

Earle, S. 1995. Sea Change: A Message of the Oceans. New York: Ballantine Books.

EPA (Environmental Protection Agency). 2002. Critical Scientific Research Needs Assessment for the Gulf of Mexico. Prepared by the Research Subcommittee of the Monitoring, Modeling, and Research Committee for the Gulf of Mexico Program Office. 47 pp.

Galtsoff, P. S. 1954. *Gulf of Mexico—Its Origin, Waters, and Marine Life.* Fishery Bulletin 89, Fish and Wildlife Service, vol. 55. Washington, D.C. 604 pp.

Kumpf, H., K. Steidinger, and K. Sherman (eds.). 1999. *The Gulf of Mexico Large Marine Ecosystem.* Malden, Mass.: Blackwell Science. 626 pp.

Tunnell, J. W., Jr. 2005. Biodiversity of the Gulf of Mexico project. Pp. 285–86 *in* P. Miloslavich and E. Klein (eds.), *Caribbean Marine Biodiversity: The Known and the Unknown.* Lancaster, Penn.: DEStech Publications.

Tunnell, J. W., Jr., and Q. R. Dokken (eds.). 2006. *Proceedings of the State of the Gulf of Mexico Summit.* Corpus Christi, Texas, 28–30 March 2006. Corpus Christi: Harte Research Institute for Gulf of Mexico Studies, Texas A&M University–Corpus Christi. 44 pp.

Tunnell, J. W., Jr., and S. A. Earle. 2004. Harte Research Institute for Gulf of Mexico Studies: Initiatives in Marine Science Research. Pp. 132–41 in R. L. Creswell (ed.), *Proceedings of 55th Annual Gulf and Caribbean Fisheries Institute,* Xel-Ha, Quintana Roo, Mexico. Fort Pierce, Fla.: Gulf and Caribbean Fisheries Institute.

Tunnell, J. W., Jr., D. L. Felder, and S. A. Earle. 2004. El Golfo de México-Pasado, presente, y futuro: Una colaboración entre Estados Unidos de América, México y Cuba. Pp. 361–71 in M. Caso, I. Pisanty, and E. Ezcurra (eds.), *Diagnóstico ambiental del Golfo de México.* Instituto Nacional de Ecología, (INECOL A.C.) and Harte Research Institute for Gulf of Mexico Studies TAMU–CC, 2 vols. México City: Secretaría de Medio Ambiente y Recursos Naturales (SEMARNAT).

U.S. Commission on Ocean Policy. 2004. *An Ocean Blueprint for the 21st Century.* Final Report. Washington, D.C.: U.S. Commission on Ocean Policy. 676 pp.

U.S. Ocean Action Plan: The Bush Administration's Response to the U.S. Commission on Ocean Policy. 2004. Washington, D.C. 39 pp.

Preface

This book is an offshoot of the highly successful State of the Gulf of Mexico Summit 2006, held at Texas A&M University–Corpus Christi, which focused on the theme of ensuring productive economies and healthy marine environments within the entire Gulf of Mexico region. The summit was designed to develop an international perspective on issues related to the Gulf of Mexico by examining the commitments and effectiveness of leaders of government, business and industry, science, and conservation and resource management. The ultimate goal was to help ensure a sustainable quality of life for future generations.

Summit participants included representatives of the governments of the United States and Mexico, along with members of regional, state, and local governments, and citizens. During the conference the most up-to-date scientific information available on the Gulf of Mexico as a large marine ecosystem was presented, with four panels sharing information about the Gulf of Mexico's economy, public health, environmental conditions, and collaborative governance.

This book focuses on information presented at the summit related to the economy of the Gulf of Mexico. While a few papers and book chapters have appeared over the years on some aspects of Gulf economics, this volume represents the first attempt to assess the Gulf as a single economic region, including both U.S. and Mexican contributions. The information provided also covers the political economy of the region. Readers will find here a collection of baseline economics data on the Gulf of Mexico against which future data can be compared to determine if the region is making progress toward the goal of achieving economic and environmental sustainability.

One missing element of the analysis is information on Cuba. It is hoped that in future years and subsequent analyses of the economic value of the Gulf of Mexico, the prevailing political situation will allow access to data and participation from Cuba.

Acknowledgments

This book would not have been possible without the convening of the State of the Gulf of Mexico Summit 2006 by the Harte Research Institute for Gulf of Mexico Studies (HRI) at Texas A&M University–Corpus Christi. Director Robert Furgason, Associate Director and HRI scientist Wes Tunnell, and conference organizer Quenton Dokken, now with the Gulf of Mexico Foundation, are thanked for their leadership, which made the summit possible. After the summit was concluded, Wes Tunnell invited the economics panel to prepare their presentations for this book, which is part of an HRI series of publications on the Gulf of Mexico. Some of the summit papers, along with one contribution invited after the conference, became the chapters in this book.

I gratefully acknowledge each of the authors and coauthors for taking the time to prepare papers for the State of the Gulf of Mexico Summit 2006, for participating in the panel presentations, and for agreeing to rework their contributions into book chapters after the summit was completed. All the authors and coauthors have full-time responsibilities in their daily professional activities, and taking on this extra responsibility indicates clear dedication to the profession and to the future of the Gulf of Mexico. Jacquelyn Whitehouse, at the University of Florida, provided administrative support for the project, and Mark Schrope, with Florida-based Open Water Media (www.openwatermedia.com), assisted the authors and me as editor, making sure the chapters would appeal to a wide audience, from economists to policy makers and citizens.

The authors also acknowledge their home institutions for allowing them the time to participate in the summit and in producing this book. This includes Louisiana State University, Monterey Bay Aquarium Research Institute, Texas A&M University–Corpus Christi, University of Florida, University of California–Los Angeles, University of Southern California, University of Southern Maine, and the Instituto de Ecología, A. C.

Introduction

This book is intended to give a broad overview of the most important factors controlling the economic significance of the Gulf of Mexico, ranging from the region's political history to assessments of the industries dependent on the Gulf. The authors represent an equally broad range and include business, economic, and ecology specialists.

Chapter 1 offers an analysis of the likelihood that the countries surrounding the Gulf will evolve in the foreseeable future into an integrated transnational community. The chapter also provides a historical foundation that supplies context for what follows. This discussion gives a valuable review of the region's fascinating and at times politically contentious history, from European colonization to the North American Free Trade Agreement. The most tumultuous events are highlighted, along with the ways these events have shaped relations among the Gulf countries—and their economies.

Ultimately, the author makes the case that each Gulf country has a stake in sustainable management of Gulf resources, and that this can only be accomplished with improved cooperation across borders. Any reasoned economic analysis of the Gulf of Mexico will conclude that the region is a critically significant component in the economies of surrounding countries. Hence it is somewhat surprising that this significance has not been adequately characterized. The remaining chapters in this book make major contributions to correcting that deficiency.

Chapter 2 addresses the questions of whether the massive economic productivity of the Gulf of Mexico can be quantified in dollars, and what the total value might be. Focusing on the economic significance to Mexico and the United States of four of the most critical and lucrative Gulf industries, the author takes a first step toward accomplishing that goal. Though value depends on many volatile factors, ensuring that it will always be in flux, the case is made that even a conservative estimate of the overall value is a significant portion of the U.S. and Mexican economies and is larger than the gross domestic product of smaller countries. The discussion also examines the potential of human activities, especially coastal population growth, to reduce that value.

Chapter 3 takes a more in-depth look at a variety of key U.S. industries that are directly or indirectly dependent on the Gulf's many natural resources. This analysis includes discussion of the extent to which various industries are dependent on the health of those resources. Beyond current values, the authors analyze trends over past decades in these industries to give a more complete view of their economic significance. They do not discuss the importance of construction and urban development in the coastal zone, which are covered in the next chapter.

Chapter 4 outlines efforts by the National Ocean Economics Program, a multi-university effort sponsored by the National Oceanic and Atmospheric

Administration, to track changes in human activities and growth patterns along the Gulf of Mexico coastline. The authors explore the region's significance as an industrial center making major contributions to the oil and gas industry and trace its rapid growth in the tourism and recreation sectors. In addition to information about longer-term trends, they examine the impact of the 2005 hurricanes on previous rates of growth along the coast in population, housing, and the economy. The authors make the case that declines observed should spur closer examination of vulnerability to such events along the Gulf Coast. Beyond more readily quantifiable economic contributions from activities examined, they also consider the non-market value generated by the Gulf ecosystem and how this value relates to quality of life for visitors and residents. This enables a more complete view of the true potential economic impacts of damage to coastal resources and ecosystems.

Finally, chapter 5 examines Mexican economic activities dependent on the Gulf of Mexico and the current level of environmental deterioration in the Mexican portion of the Gulf. As pointed out, Mexican Gulf states make relatively low contributions to Mexico's gross domestic product, and the authors suggest that this may be due in large part to decreases in the integrity of the Gulf of Mexico ecosystem. They argue that continued unsustainable development will likely lead to crisis in the region. The authors conclude that to prevent this crisis, present and future coastal planning efforts must do a better job of considering environmental concerns, including climate change impacts.

The studies presented here make a clear case both quantitatively and qualitatively that the Gulf of Mexico's resources are a vital component in the economies of the United States, Mexico, and Cuba. Though this book provides a more thorough review of the economic significance of those resources than has ever been assembled, it remains only a starting point for complete understanding of the Gulf's importance.

The hope of all involved is that this effort will indeed be a starting point for further research. We also hope that the information presented will be a valuable resource not only to researchers but also to the public and the resource managers who must determine the degree to which the health of the Gulf ecosystems will be considered as coastal populations continue to grow and increased pressure is placed on Gulf resources.

Gulf of Mexico Origin, Waters, and Biota

Volume 2, Ocean and Coastal Economy

1

The Gulf of Mexico Region as a Transnational Community

TERRY L. MCCOY

Introduction

This chapter assesses the prospects for the Gulf of Mexico region to evolve into an integrated transnational community. The underlying question is whether the Gulf functions as a barrier separating or a bridge uniting the coastal regions of the three countries that share it. Answering that question involves addressing a number of related ideas: Are trade and investment flows, transportation networks, demographic movements, intergovernmental collaboration, and civil society interaction knitting the Gulf territories of the United States and Mexico together across the Gulf? Do officials and residents of the region think of themselves as belonging to a Gulf community? Is there a growing sense of community identification accompanied by transnational institution building? And where does Cuba, the third Gulf nation, fit?

The original impetus for this research, which began in the mid-1990s, was the launch of the North American Free Trade Agreement (NAFTA), which promised a new era in United States–Mexico relations (see McCoy, 1996, for early work). A decade later it is appropriate to assess the extent to which the predicted changes have in fact occurred.

In the early 1990s the negotiations leading up to NAFTA were followed by the battle for congressional approval in the United States. This debate focused attention on the special relationship between the United States and Mexico, two North American nations with different historical origins and different levels of economic development. To an important extent, the long land border shared by the two countries had been the defining geographic linkage in the bilateral relationship. The southwestern United States and northern Mexico–linked cross-border trade and investment, tourism, migration, and other activities constituted an authentic "border region" (Raat, 1992). But the United States and Mexico shared another border—that defined by the Gulf of Mexico—and to a lesser extent NAFTA refocused attention on the Gulf region. At the 1995 Gulf of Mexico Symposium, a Texas state official referred to the Gulf as the "forgotten border" between the United States and Mexico, according to a *Miami Herald* article. The agreement calling for formation of an association of the Gulf states of Mexico and the United States emphasized that they share a "well-defined geographic region" and "many areas of common interest" (Accord of the States of the Gulf of Mexico, 1995).

To examine the extent to which the Gulf is now, or in the post-NAFTA era is evolving into, an integrated transnational border region or community analogous to the land frontier shared by the United States and Mexico, this chapter begins by offering a working definition of the Gulf of Mexico region. It then reviews the history of relations in the Gulf area. In this regard, the historical record provides little evidence that prior to NAFTA the Gulf in fact constituted an integrated transnational region defined by clear separation from surrounding territory and interdependence of the coastal territories that make it up (Bassols Batalla Angel in Ávila Sánchez, 1993). Subsequent analysis of the region's current configuration, paying particular attention to economic flows across the Gulf and intergovernmental cooperation following implementation of NAFTA, provides scant evidence that the Gulf of Mexico is evolving toward becoming a transnational community. An obvious obstacle to a fully integrated Gulf is the anomalous status of Cuba, which has normal bilateral relations with Mexico but has not had full diplomatic and trade relations with its Gulf neighbor to the north for over four decades.

Defining the Gulf Region

The Gulf of Mexico is "a Mediterranean-type basin bounded by the North American continent and the island of Cuba" and "connected with the Caribbean Sea by the Yucatán Strait and with the Atlantic by the Straits of Florida" (Darnell and Defenbaugh, 1990). Making up the region's land territory are eleven maritime states of the United States and Mexico (Florida, Alabama, Mississippi, Louisiana, and Texas; Tamaulipas, Veracruz, Tabasco, Campeche, Yucatán, and Quintana Roo) plus the four Gulf Coast provinces of Cuba (Pinar del Río, Havana, Havana City, and Matanzas).

The Gulf The Gulf of Mexico itself is the defining geographic feature of the region. The Gulf's three major physiographic features are "a shoreline bordered by coastal plains and mountains; a surrounding, generally wide, continental shelf; and a large, off-central basin" (Gore, 1992). The Gulf is one of the world's largest and deepest marine basins. Its maximum width is approximately 1,000 miles in an east-west direction, and its narrowest width is about 500 miles in a north-south direction, from the Mississippi delta to the Yucatán Peninsula. The marine continental shoreline, from Cape Sable, Florida, to the tip of the Yucatán Peninsula, is about 3,600 miles, with the Gulf coast of Cuba adding another 240 miles in length. But the total U.S. Gulf shoreline alone, with bays, inlets, and other features included, is over 17,000 miles long (Gore, 1992).

The major natural resources of the Gulf of Mexico are its marine life and petroleum deposits, both of which have played a role in shaping the history and economics of the region. The Gulf also supports waterborne transportation and recreation. The important features of the Gulf affecting shipping are its winds, which vary considerably with location and season; surface currents, the most important being the Yucatán Current and the Loop Current; and storms (Sheppard, 1994). Its waters and surrounding territory experience seasonal tropical storms and hurricanes.

The 1982 United Nations Law of the Sea Treaty gave the United States and Mexico bigger stakes and greater control over the resources of the Gulf. Ninety-five percent of its harvestable living resources, 98% of the petroleum resources, and significant mineral resources lie within each country's 200-mile economic zone in the Gulf, authorized by the treaty (Gore, 1992).

The Gulf, rich in marine life, has historically been an important source of fish, seafood, employment, and income for the three countries that border it as well as for the other nations that fish it. In addition to commercial fishing, the Gulf supports growing recreational fishing, diving, and marine-based tourism—all of which are increasingly in conflict with commercial fishing—in the United States, Mexico, and Cuba, as well as a declining artisanal industry in Mexico. Pressures to increase fish catches, along with overlapping state, national, and international jurisdictions over Gulf waters and seafloor, put marine life at risk of overexploitation.

Petroleum constitutes the Gulf's other important natural resource, and one not easily compatible with the Gulf's marine life. The first successful U.S. offshore oil drilling occurred in 1938 off the Louisiana coast (Gore, 1992). From there, oil exploration in the Gulf expanded rapidly, and on the eve of NAFTA, it was the focal point of U.S. offshore activity. In 1994 the Gulf of Mexico accounted for "91 percent of the total acreage under lease in the Alaska, Gulf of Mexico, and Pacific OCS [Outer Continental Shelf] regions and . . . over 76 percent of the bonus high [lease] bids" (MMS, 1994). With 810 active fields, the Gulf produced 89% of all OCS oil and 99% of OCS gas. In 1993, estimated reserves in fields leased by the Minerals Management Service stood at 2.144 billion barrels for oil and 29.090 trillion cubic feet for gas (MMS, 1994). Intense bidding for Gulf leases in 1995 confirmed that the Gulf was likely to remain the most important petroleum region in the United States for some time (Salpukas, 1995).

Although the Mexican oil industry has a long history along the Gulf Coast, the Mexican state oil company PEMEX did not drill the first offshore wells until the 1960s. Following the Mexican oil boom of the 1970s, the head of PEMEX described the Campeche Sound area off the west coast of Yucatán as "the greatest offshore field ever" (quoted in Grayson, 1980).

The steep increase in global energy prices experienced early in the twenty-first century triggered renewed interest in deepwater exploration in the Gulf and put added pressure on the United States and Mexico to open their Gulf waters to drilling. In early 2006, prospects for new oil discoveries, plus active Cuban exploration in its territorial waters, led the U.S. Congress to relax drilling restrictions in the Gulf. Then, in September 2006, three large multinational oil companies announced that their deepwater test wells in U.S. Gulf waters had made a major discovery, which "confirms very large reserves of recoverable oil in the Gulf" (Daniel Yergin of Cambridge Energy Research Associates, quoted in Krauss, 2006).

Gulf States and Provinces　The Gulf states are part of a great coastal plain that stretches from the southeastern Atlantic Coast of the United States to the Yucatán Peninsula of Mexico. This plain is bounded by the Appalachian chain on the north in the eastern United States, bisected by the Mississippi River system, and bounded on the west by the Sierra Madre Oriental in Mexico. With a very gentle slope, the plain is broad in the United States and northern Mexico but nearly disappears in the mountainous state of Veracruz.

The natural resources of the Gulf states are their climate, soils, rivers, and minerals. The onshore petroleum deposits of Louisiana, Texas, and the Mexican Gulf Coast are the most important mineral resources. Non-petroleum minerals include the phosphate rock of Florida; salt (halite) in Louisiana and Texas; and sand, clay, gravel, and sulfur from various areas (Gore, 1992). The region's climate and soils constitute important agricultural resources. Florida, and to a lesser extent Louisiana and Texas, produce citrus, cane sugar, winter fruits, and vegetables. This combination is found on Mexico's Gulf Coast, while the middle Gulf states of Alabama and Mississippi specialize in temperate grains, cotton, livestock, and poultry. Forests have been an enduring resource on the Gulf Coast, while the estuaries are ideal for aquaculture. The Gulf and the region's rivers constitute a major transportation resource for the United States, and together with the climate they have supported the development of tourism in the twentieth century (Gore, 1992).

Despite the Gulf region's substantial resources, most notably petroleum, the Gulf states in the United States and Mexico have historically constituted poor, peripheral regions for both countries. The same is not true of Cuba. The location of Havana—Cuba's capital city, principal port, and economic and population center—endows Cuba with a strong Gulf orientation, more so than was historically the case for either the United States or Mexico.

History of Transnational Relations on the Gulf

Although a common, harmonious history is not a prerequisite for regional integration, as post–World War II Europe demonstrates, such a history can help shape a modern transnational community. However, the history of the Gulf of Mexico following European discovery was anything but harmonious. Competition among the leading imperial powers left the Gulf fragmented, not unified, after three centuries of European colonization. By the end of the nineteenth century, the pattern emerging in relations among the three independent nation-states that shared the Gulf was one of asymmetry under U.S. domination.

European Colonization   The Gulf region was one of the earliest sites of European contact in the Americas, but European conquest and colonization proved disastrous for the native peoples and cultures. Of the peoples inhabiting the entire region at the time of conquest, only the Mayas of the Yucatán Peninsula and highland Indians of Mexico survived. While successful in eliminating indigenous threats to their colonization projects, the Europeans did not succeed in imposing unity on the Gulf.

Columbus reached Cuba on his first voyage, and the Spanish returned for good in 1511. From Havana, they set out to conquer and colonize the territory of the Gulf. In 1517 the governor of Cuba sent an expedition to explore the Gulf Coast of Mexico. When it returned to Cuba with stories of wealthy Indian civilization, he dispatched Hernando Cortés on another expedition. Cortés established the Gulf settlement of Veracruz, from which he launched the well-chronicled conquest of the Aztec empire from 1519 to 1520 (Burkholder and Johnson, 1994). The Spanish experience in Florida was less rewarding. Several expeditions—including Hernando de Soto's from 1539 to 1542—explored the peninsula from Havana, but settlement did not occur until later because Florida had little apparent wealth. West Florida, a separate subunit in the Spanish empire, was settled from Mexico and administered through Veracruz during the first Spanish period.

Although late settlement was the pattern around the Gulf, the Spanish presence and trade routes were established by the end of the sixteenth century (Gore, 1992), and it did not take long for other European powers to be attracted to the Gulf. To protect its trade from both private buccaneers and foreign navies, Spain instituted a fortified fleet system to carry treasure back to the mother country. One of the three major routes was between Veracruz and Seville through Havana, which claimed for Cuba the title of "Llave del Golfo," or Key to the Gulf. As a result Havana emerged as the dominant city in Cuba, instead of Santiago on the southeastern coast, and Cuba developed a predominantly Gulf orientation rather than a focus on the Caribbean (CubaNews, 1995). The Spanish founded St. Augustine on the northeastern frontier of its incipient New World empire in 1565 and incorporated its fortifications into the strategic defense system centered in Havana. Florida functioned as an appendage of Mexico and Cuba, subsidized by a stipend (*situado*) from the viceroyalty of New Spain funneled through Havana (Bretos, 1991).

France and England soon became Spain's principal rivals on the Gulf, with the primary French challenge coming from the north. Their earliest thrust, which the Spanish repelled, was in northeast Florida. Then in 1682 LaSalle descended the Mississippi from Canada and claimed the entire Mississippi delta for France. LaSalle's challenge, and the French maps of the region, which fell into Spanish hands, heightened Spain's interest in the Gulf Coast (Jackson et al., 1990). The British challenged Spanish hegemony from both south and north. From the north, Britain made repeated forays by land and sea into Florida, while they also threatened Spanish territory from the Caribbean.

Spain's response to threats from European rivals was to secure the northern boundary of its American empire. As buffers against French threats, the strategy included building a network of missions in Texas beginning in the 1720s and establishing Fort San Carlos on Pensacola Bay on the northern coast; the fort was recaptured in 1723 after four years of French occupation. Spain successfully defended St. Augustine against the English in the wars for Spanish Succession, and successfully defended Jenkins's Ear on the northeastern border of its defensive perimeter (Burkholder and Johnson, 1994).

In spite of these moves, Spain's position in the Gulf region gradually eroded, and the Gulf of Mexico became increasingly fragmented. In the French and Indian War from 1754 to 1763, in which Spain sided with France, the British captured Havana and then forced Spain to accept it in exchange for Florida in the peace settlement. France conceded most of its North American territory to Great Britain in this same settlement (Gannon, 1993). By siding with the rebels and engaging the British militarily along the Gulf Coast, Spain regained Florida in 1783 following the independence of the United States from Great Britain. But the country could not long withstand the expansionist pressures of the new American nation (Gannon, 1993). In 1819 Spain and the United States signed the Adams-Onis treaty transferring Florida, which was then two colonies, to the United States.

Spain's loss of Florida was followed by its defeat at the hands of Mexicans fighting for their independence. With this, Spain's Gulf presence was reduced to its colony in Cuba, and the United States steadily eclipsed Spain as the dominant Gulf power.

U.S. Expansion in the Nineteenth Century   A defining feature of relations in the Gulf is the hegemony—and ongoing resistance to it—of the United States. On the eve of its independence in 1823, Mexico had been Spain's richest, largest, and most important New World colony, but the first decades as an independent republic would prove difficult for Mexico. At the time the United States was a young agrarian country and Cuba was a forgotten Spanish colony. Over the rest of the nineteenth century, the United States gathered strength economically and militarily, while Mexico and Cuba stagnated, and this growing disparity facilitated U.S. domination of the Gulf region.

Even before acquiring Florida, the newly independent United States had purchased the Louisiana Territory, securing its first major outlet to the Gulf, while at the same time eliminating France from the region and denying additional territory on the Gulf to Spain (Gore, 1992). Louisiana achieved statehood in 1812, which was the same year Andrew Jackson defeated the British in New

Orleans in the War of 1812 to end the English threat on the Gulf for good; statehood followed for Mississippi in 1817 and Alabama in 1819. Invoking the geopolitical spirit of the soon-to-be-pronounced Monroe Doctrine, the United States acquired Florida in 1819 to prevent it from being transferred from Spain to Great Britain (Langley, 1989), and this extensive Gulf territory became a U.S. state in 1845. The annexation of Texas, also in 1845, was the final building block in the U.S. Gulf presence.

Agriculture dominated the five Gulf States, with the primary export crop, cotton, shipped through the port of New Orleans. Acquisition of Louisiana linked the northern United States to the Gulf of Mexico via the Mississippi River, which, along with the Ohio and Missouri rivers, created a water-based transportation system connecting the interior of the country to New Orleans, the Gulf, and the world (Gore, 1992).

Statehood for Texas was the direct result of the military defeat of Mexico. This event inevitably affected subsequent relations between these two Gulf neighbors, and as Raat (1992) puts it, "signified the end to any likelihood that Mexico, rather than the former thirteen colonies of British North America, would become the predominant power in North America."

The loss of Mexican territory to the United States in the war of 1845 to 1848 was the culmination of two converging forces—Spain's and later Mexico's neglect of what was then northern Mexico, and the westward expansion of the United States. Loose Mexican control, heavy American immigration, and growing commercial ties meant the de facto incorporation of northern Mexico into the American Southwest. Even during the 1830s, the state of Yucatán was linked more closely with New Orleans and New York than with central Mexico (Raat, 1992). Under the 1848 U.S.-dictated Treaty of Guadalupe Hidalgo, Mexico formally conceded half its territory to the United States, including not only Texas and its long Gulf shore but also the huge California and New Mexico territories. In 1853 Washington acquired another 30,000 square miles of Mexican territory through the Gadsden Purchase, consummated by President Santa Anna in a desperate attempt to raise funds for his bankrupt government (Meyer and Sherman, 1991).

The Civil War interrupted westward expansion of the United States and weakened the U.S. presence on the Gulf during the conflict. All five Gulf States joined the Confederacy, and Gulf shipping and ports became contested military targets. On the Gulf Coast of Florida, for example, Union forces held key fortifications in Key West, the Dry Tortugas, and Pensacola Bay. Mexico, however, was unable to take advantage of the war to reassert itself on the Gulf, although the port of Matamoros, just south of the U.S. border, became a transshipment point for Mexicans doing business with Confederates in Texas (Raat, 1992). In 1861 Spain, Britain, and France landed troops in Veracruz with the announced mission of collecting Mexican customs receipts to repay European creditors, but Napoleon III deployed the French troops to install Austrian archduke Maximilian as emperor of Mexico in 1863. His rule was brief, however. It was in part Washington's threat, as the Civil War was ending, to invoke the Monroe Doctrine that forced Maximilian and the French to abandon Mexico in 1867 (Meyer and Sherman, 1991). It also presaged a return of the United States to the Gulf.

In the first two-thirds of the nineteenth century, the relationship between the newly independent North American republics of the United States and Mexico quickly became marked by conflict and asymmetry. Mexico suffered losses in territory and national honor to the "Colossus of the North." And it fell farther behind economically. Half as productive as the United States in 1800, Mexico fell to only one-eighth of U.S. productivity in 1867 (Raat, 1992). The developed/ underdeveloped dichotomy between the two large nations sharing a 2,000-mile border was a distinguishing feature of U.S.-Mexican relations by 1875 and remains so today.

With the end of the U.S. Civil War and the defeat of Maximilian, the United States and Mexico entered a new era. For the United States, it brought industrialization and modernization at home and commercial expansion and political adventurism south of the border. For Mexico, it was a struggle to catch up, a struggle that increased the country's dependency on the United States. Langley (1989) characterizes this as an example of core/periphery relations in which "the United States sought markets, bases, and influence. Mexico, like most of Latin America, adjusted because the requirements of modernization seemed to dictate doing so."

Although the drama of the last quarter of the nineteenth century played out largely along Mexico's northern land frontier known as the Gran Chichimeca, the Gulf coasts of both countries experienced change. For the United States, struggling to recover from the Civil War and the end of slavery, change came in the form of railroads, coal and iron industries, commercial forestry, cattle raising, oil, and the first stirrings of citrus and tourism in the most economically underdeveloped of the five Gulf States, Florida (Gore, 1992). The Mexican Gulf Coast followed a similar path.

Under the dictatorship of Porfirio Díaz (1876–1910), Mexico experienced growth and modernization with heavy U.S. participation. The development model during Díaz's reign, known as the *Porfiriato*, favored the imported capital, technology, and entrepreneurial skills attracted by friendly investment laws, generous land grants, and imposed political stability. By 1911 foreigners controlled over 20% of the land in Mexico and the most important nascent industries. U.S. investment, concentrated in railroads, mining, and real estate, represented nearly 38% of all foreign capital (Raat, 1992).

Railroad construction was a major force for change in Mexico, as in the United States. The major lines, which ran north and south, linked Mexico City to the United States and transformed Mexico's northern states from a frontier to a border, "a place in which U.S. capital goods greeted Mexican labor under rules created by politicians and their foreign allies in Mexico City." By 1900, "the northern borderland was the most modern section of Mexico" (Raat, 1992). On the Gulf, Meyer and Sherman (1991) argue that completion of the Mexico City–Veracruz railroad in 1872 helped initiate the modernization carried out during the *Porfiriato*. Gulf ports were subsequently modernized. By the turn of the century, there were ten serviceable Gulf ports and several bustling port cities (Meyer and Sherman, 1991). The railroads and ports linked the henequén fiber producers of the Yucatán and coffee growers of Chiapas to new export markets, while at the same time creating the uneven patterns of development characteristic of postbellum plantation agriculture in the American South (Raat, 1992). Finally, it was on the Gulf Coast that Mexico's oil industry began to take shape under concessions granted to American and British investors (Grayson, 1980).

In the eastern Gulf, Cuba, then still a Spanish colony, began to stir economically during the nineteenth century with the emergence of a modern sugar industry (Burkholder and Johnson, 1994). As the century progressed, the United States and Cuba drew closer across the Gulf through U.S. investments and Cuban migration. Once again Havana's location on the Gulf strengthened the island's status as a Gulf territory (CubaNews, 1995). Cubans moved back and forth to work in the cigar industries of Key West and Tampa, Florida. In 1886 there was steamship service between Havana and Tampa via Key West three times per week, and by 1900, 20% of Tampa's population was Cuban (Bretos, 1991). Similar ties existed with the Gulf Coast of Mexico, especially between the Yucatán and Cuba. Novel (1994) reported, "Until well into the twentieth century, it was much easier to travel to Cuba from Yucatán . . . than go to the capital of the republic. In these terms, Yucatán served as a point of contact between Mexico, the rest of the Caribbean, and coastal zones of the Gulf of Mexico, rather than a dividing line or barrier."

By the end of the nineteenth century the Gulf of Mexico had become a very different place. With Cuba's independence from Spain in 1898, the three Gulf nations were nominally sovereign republics. During the latter part of the nineteenth century, the United States and Mexico, and to a lesser extent Cuba, had experienced significant economic development and social change as well. The most striking feature of the region, however, was the emergence of the United States as the dominant power. If by 1900 the "dependency of Mexico and the modernized Mexican borderlands on the United States was greater than ever," as Raat asserts (1992), U.S. hegemony was equally pronounced on the peripheral territories of the Gulf, including Cuba. In the case of Cuba, the Gulf was inescapably central to the relationship with the United States. It was part of the Cuba-Mexico relationship as well, but the land border dominated U.S.-Mexico relations.

Relations in the Early Twentieth Century   In the first three decades of the twentieth century, the United States actively continued to dominate the Gulf of Mexico, though with the Great Depression and then World War II, the Gulf lost some of its strategic significance to the United States. The Cuban revolution of 1959 refocused U.S. attention on the Gulf, while the emergence of the New South in the 1960s in the United States and the 1970s Mexican petroleum boom elevated the Gulf region to greater importance within each country. But the rupture in U.S.-Cuba relations disrupted relations across the Gulf.

Following the Spanish-American War at the beginning of the twentieth century, the United States aggressively expanded its economic, strategic, and cultural presence throughout Latin America, giving special attention to the Caribbean basin. Relations with Cuba and Mexico were central components of the foreign policies of American administrations from 1898 to 1932 and were the source of considerable tension with the two Gulf neighbors. After defeating Spain, U.S. forces occupied Cuba from 1898 until 1902. Washington imposed on the 1901 Cuban constitution the Platt Amendment, which incorporated the right of U.S. intervention and conceded the Guantánamo naval base to the United States as the price of recognizing the island's independence. "A sovereign nation in name, Cuba was in fact a protectorate of the United States," observes Peter Smith (2000). Washington intervened under cover of the Platt Amendment on several occasions in an attempt to bring order to Cuban politics. U.S. investors dominated Cuban trade and commerce. The Cuban economy was also tied to the U.S. market through the sugar quota and other preferential trade agreements. Following independence from Spain, many Cubans living in Florida went to their homeland, only to return subsequently to the United States as the situation on the island periodically deteriorated. Tampa, Key West, and Miami all developed significant Cuban expatriate communities with strong ties across the Gulf to the island. The advent of aviation facilitated transportation and communication, and two-way tourism developed between Cuba and Florida (Bretos, 1991).

Although Mexico never became a U.S. protectorate, Washington did intervene and interfere in Mexican affairs. Its principal preoccupation was the 1910 to 1917 Mexican revolution. With its newly created sphere of interest in the Caribbean basin, and extensive private economic interests in Mexico, the United States found the upheaval and unrest of the revolution to be unacceptable. In 1914 President Woodrow Wilson dispatched U.S. forces to occupy the port of Veracruz. In 1916 he sent an expeditionary force across the land border into northern Mexico. Beyond these attempts to shape the course of the revolution, Washington directed efforts at heading off implementation of the radical reforms called for in the Constitution of 1917. U.S. and other foreign oil companies feared that Article 27 of the constitution, which gave the government all rights to the country's subsoil resources, would be invoked to expropriate their holdings. In 1923, in return for U.S. diplomatic recognition of the revolutionary regime, Mexico agreed not to apply Article 27 retroactively (Raat, 1992), but the issue later reemerged.

The revolution was cataclysmic, taking between 1.5 and 2 million lives and inflicting extensive property damage in Mexico. Though nationalism with a strong anti-American orientation was a rallying cry for revolutionary forces, the revolution actually strengthened the complex web of Mexican dependency on the United States. Tens of thousands of Mexicans from all social classes crossed the U.S. border as refugees from the revolution (Meyer and Sherman, 1991). Once the violence subsided, many, but not all, Mexicans returned to their home country with new economic and family ties to the United States. Through them, the United States had new links to Mexico. U.S. merchants along both the northern border and the U.S. Gulf Coast did business with the armies of the revolution, and U.S. investors, geologists, engineers, and skilled laborers helped produce a petroleum boom on Mexico's Gulf Coast in the midst of the revolution. By 1921 Mexico accounted for 21% of world oil output (Raat, 1992), thanks in part to the U.S. occupation of Veracruz and naval vessels patrolling the Gulf, which helped protect the oil zone from the revolutionary upheaval (Grayson, 1980).

Good Neighbor to Cold Warrior Two events dominated transnational relations in the Gulf of Mexico in the first third of the twentieth century: U.S expansion and the Mexican revolution. During this period, the U.S.-Mexican relationship grew quite tense, as did Washington's relations with Cuba. Then in the early 1930s, growing Latin American hostility toward U.S. policies combined with the deepening of the Great Depression and the growing threat of another world war to lead the United States to step back from intervening in the affairs of its neighbors. Franklin Roosevelt's Good Neighbor policy promised Latin America "nonintervention, noninterference, and reciprocity" (Langley, 1989). For the Gulf region, this meant a less aggressive U.S. posture.

A major test of the new relationship between Mexico and the United States occurred in 1938 when President Lázaro Cárdenas nationalized the Mexican oil industry, which had been dominated by U.S. and European companies. Instead of responding with action to force the Mexican government to reverse its decision, as Washington might have done in the past, the Roosevelt administration pursued a negotiated settlement that recognized Mexico's right to nationalize its oil industry as long as it compensated the affected companies. With peaceful resolution of the oil dispute in 1941, and the outbreak of World War II, Mexico and the United States enjoyed their "closest economic, political, and military ties ever," observed Grayson (1980). Mexico supplied important raw materials and cheap labor for the U.S. war effort, while the leaders of the "institutional revolution" steered a more moderate path that carried over into the postwar period, when successive Mexican presidents pursued a strategy of import-substitution industrialization. U.S. companies welcomed the stable investment climate, and Mexico became a popular destination for U.S. tourists. The postwar Mexican "economic miracle" produced high growth rates, although U.S. per capita GDP was still more than three times that of Mexico (Raat, 1992).

Repeal of the Platt Amendment in 1934 ended direct U.S intervention in Cuba, but the U.S. government continued to meddle in Cuba's volatile internal politics. Cuba increased its dependence on the American sugar market, and in 1934 the two countries signed a new reciprocal trade treaty (Pérez, 1988). Two-way tourism grew during the 1930s. As with Mexico, World War II produced an economic boom in Cuba, but the island turned increasingly chaotic in the postwar years. In 1959 growing political grievances and deepening socioeconomic frustration brought to power the leftist Fidel Castro regime, which soon began Latin America's second major social revolution, one with deep anti-American roots.

The impact of the Cuban revolution on the transnational relations of the Gulf was direct and dramatic. With a self-declared Marxist-Leninist leader aligned with the Soviet Union ruling Cuba, the Gulf became for the United States a frontline zone in the Cold War. Within two years of Castro's coming to power, the United States and Cuba severed diplomatic relations. On the heels of this break, Washington took steps to enact a comprehensive economic embargo, effectively suspending trade, investment, and tourism across the Straits of Florida, as part of a policy to contain Cuba's communist revolution and undermine the Castro regime. The two Gulf nations—once the closest of neighbors—became the "closest of enemies" (Smith, 1987) and have stayed so through five decades. Hundreds of thousands of Cubans fled to the United States and settled in South Florida, just 90 miles from Havana. The United States and Cuba were destined to stay locked in a special relationship across the Gulf, even though the two countries still do not conduct normal diplomatic and economic relations.[1]

Of the Latin American countries, Mexico alone resisted U.S. pressure to break relations with and adopt economic sanctions against Cuba, and Mexican policy on Cuba served as reaffirmation of the country's independence from Washington. Sagebien and Tsourtouras (1999) also attribute Mexico's insistence on maintaining full relations with Cuba to "the fact that both countries share a revolutionary history and to the application to relations with Cuba of the core principle of Mexican foreign policy—non-intervention." But under the embargo, Mexico benefited commercially from the continued cutoff of Cuban access to U.S. markets and investment, and Mexican authorities also quietly shared intelligence information with their U.S. counterparts on travelers going to and from Cuba through Mexico (Raat, 1992). By maintaining diplomatic and trade relations with Cuba, Mexico kept Cuba from being completely frozen out of the Gulf, although the Cuban revolution and the U.S. response to it altered the Gulf region substantially.

Boom and Bust In the 1960s the U.S. Gulf states joined the other Sunbelt states in a cycle of rapid economic and demographic growth. The transformation was most dramatic in Florida and Texas, but it touched the whole region.

The Mexican Gulf region also experienced development, but the results were less sustained than for the United States. The motor of change in Mexico was petroleum. In the mid-1970s, just as world oil prices were skyrocketing, the Mexican state oil monopoly, PEMEX, discovered large new oil reserves in the states of Tabasco and Chiapas and offshore in the Bay of Campeche. Although the announced intention of President José López Portillo, who assumed office in 1976, was to bring the new fields on line slowly in order to avoid disrupting the economy and to conserve petroleum as a future asset, Mexico was in a deep economic crisis, and exporting oil seemed like the easiest way out of it (Grayson, 1980). Bringing the new oil fields into production would also give Mexico leverage in its relations with the United States. For these reasons, the López Portillo administration launched a rapid expansion of oil exploration and production. By 1981 Mexico's verified reserves were second only to those of Saudi Arabia, and the country experienced a heady, if short-lived, petroleum-led boom.

During the late 1970s the country was awash in international loans to finance its economic expansion. It was Mexico's subsequent default in 1982 on its foreign debt that triggered the country's severe decade-long economic crisis that rapidly spread throughout Latin America to create a regionwide debt crisis and deep economic recession.

The petroleum bonanza turned out to be a mixed blessing for the Gulf region of Mexico, just as it did for the country as a whole. Although most of the oil and gas discoveries were in the Gulf of Mexico, the region did not receive a proportionate share of the benefits from their exploitation. Federal spending in the region did increase following the discovery of the new fields, and regional per capita income grew faster than the national average from 1970 to 1980. But local development was uneven, reflecting a national priority of increasing oil production rather than a commitment to balanced regional development (Randall, 1989). The prospects for new jobs in the oil industry attracted migrants from throughout Mexico and strengthened the hand of the powerful oil workers' union run out of Tampico. Within the region, new centers of economic power grew up in southern areas such as Coatzacoalcos, Villahermosa, and Ciudad del Carmen, somewhat diluting the traditional power of Veracruz (Riding, 1984). Overall, the petroleum boom exacted a toll on the Gulf region, according to Randall (1989): "Agriculture shrank. The polluting effects of the oil industry damaged agriculture, cattle raising, and fishing interests. Food availability and diet worsened. The immigration increased land prices, and put a burden on schools, hospitals and other social services. Prostitution flourished."

In June 1981 world oil prices plummeted, and the Mexican petroleum boom came to an abrupt end. In August of the following year Mexico declared that it was no longer able to service its foreign debt, and the country entered its worst economic downturn since the Great Depression. The crisis was especially painful in the Gulf region. Between 1980 and 1990, four of the six Gulf states actually experienced a decline in population. The crisis, combined with external pressures, led the successors of López Portillo to embrace a comprehensive economic reform package of which the North American Free Trade Agreement turned out to be the centerpiece.

The Gulf of Mexico Region in the NAFTA Era

Prior to the NAFTA negotiations, there was little in the history of relations on the Gulf to suggest that the region had the potential to become an integrated transnational community. To the contrary, with the exception of the early Spanish colonial period, there had been few sustained instances of political integration of the Gulf territories. Since the early nineteenth century, interaction among three Gulf nations had been highly asymmetrical and often unfriendly. Cuba and the United States had not conducted normal state-to-state or trade relations since the early 1960s. Mexico and the United States had had a diplomatically correct but frequently chilly relationship. Among other things, Washington objected to Mexico's refusal to follow the U.S. lead and break relations with Cuba. The two neighbors also disagreed over how to resolve the armed insurrections raging in Central America in the 1980s.

On the Eve of NAFTA   Beyond the deliberately circumscribed nature of official relations, economic and demographic ties had created complex webs of linkages among the three Gulf neighbors. If the paradoxes of U.S.-Cuba relations made for the closest of enemies, then Mexicans could be described as "distant neighbors" of the United States.[2] However, in the early 1990s, changes were under way in Mexico that would draw it closer to the United States and open up the possibility of integration of the U.S. and Mexican Gulf states into a transnational community—or at least a marine border zone comparable to the land frontier of northern Mexico and the southwestern United States.

NAFTA was both a cause and an effect of the profound transformation that swept through Mexico in the 1990s. The country entered the decade governed by the one-party political regime that had co-opted the Mexican revolution and had erected trade and investment barriers to pursue an inward-looking, protected development strategy. By the end of the 1990s, Mexico had shifted to an open, market-oriented political economy and, with the victory of an opposition candidate in the July 2000 presidential election, a competitive multiparty political system.

The centerpiece of Mexico's new political economy—necessitated by the economic collapse of the 1980s and actively promoted by President Carlos Salinas—was the 1994 free trade agreement with the United States and Canada, which "aimed to eliminate all trade and investment barriers and level the playing field on procurement, telecommunications, banking services and other sectors" (Pastor, 2005a). It also incorporated a state-of-the-art dispute-settlement mechanism but did not erect an institutional infrastructure because the agreement was designed to be self-executing. NAFTA reflected the reality of firm-level border/cross-border integration already under way under the auspices of the *maquiladora* program, which had paved the way for U.S. companies to set up operations in northern Mexico. But by creating binding rules and procedures for free trade and investment flows among the three members, NAFTA opened the Mexican economy to much deeper integration with the United States, and in doing so signaled an end to "the long-standing Mexican tradition of keeping relations at an arm's length from the dangerous northern neighbor" (Dominguez and Fernandez de Castro, 2001). In other words, Mexico was casting its lot with the United States. In 2001 George W. Bush also broke tradition by making his first official trip outside the country as president to Mexico, where he and President Vicente Fox pledged that their governments would give highest priority to reshaping bilateral relations. On welcoming President Fox to Washington on September 5, 2001, President Bush declared, "We are building a relationship that is unique in the world, a relationship of unprecedented closeness and cooperation" (quoted in Bondi, 2004). For its part, the Fox administration heralded NAFTA as the important first step in creating a North American community along the lines of the European Union (Bondi, 2004).

Just as NAFTA marked the start of a new era in U.S.-Mexico relations, and with it the possibility of closer integration of the U.S. and Mexican Gulf states, U.S.-Cuba relations appeared to be at a crossroads at the beginning of the 1990s. The end of the Cold War and collapse of the Soviet Union raised expectations of political and economic changes in Cuba that would be even more far reaching than those occurring in Mexico. A U.S.-Cuba rapprochement paving the way for reincorporation of Cuba into the Gulf community also seemed to be in the cards. Now, more than a decade after NAFTA went into effect, it is possible to assess changes in relations among the three Gulf nations and their consequences for relations around the Gulf.

Trade and Investment The one undeniable outcome of NAFTA has been dramatic increases in trade and investment flows between Mexico and the United States, which were the principle goals of the agreement. Mexican tariffs, which averaged over 10% and were 2.5 times U.S. rates, have fallen in keeping with NAFTA schedules, and most non-tariff barriers have been eliminated. U.S. exports to Mexico quadrupled from $28 billion in 1993 to $111 billion in 2004, while U.S. direct investment in Mexico grew from $1.3 billion in 1992 to $15 billion in 2001 (Pastor, 2005a). Mexico became the United States' second leading trading partner (after Canada), and approximately 90% of Mexican exports, which are predominantly manufactured items, go to the United States. Incorporating Canada into the analysis shows that the share of intraregional North American exports grew from 30% of total exports in 1982 to 58% in 2002 (Pastor, 2005a). Since much of this trade is either intraindustry or intra-firm trade, many industries and firms are assuming a genuinely North American character. Remittances from the millions of Mexicans working in the United States back to their families in Mexico constitute another very important pillar of the expanding North American economy. The increase in cross-border trade, investment, and remittances has not diminished the development gap between the United States (and Canada) and Mexico. Mexican GDP per capita is approximately only one-sixth that of the United States.

One of the assumptions provoking reconsideration of the Gulf of Mexico was that the predicted jump in U.S.-Mexico trade under NAFTA would overwhelm the land border and generate greater use of the Gulf "superhighway" for waterborne commerce. Although trade across the Gulf has increased at a rate comparable to the overall increase in U.S.-Mexico trade, it has not increased its relative share of trade between the two countries, which remains at about 6% (Varney, 2005; confirmed in March 2006 interview with Gary Springer, secretary general of Gulf of Mexico States Accord). In spite of the presumed cost advantage of Gulf shipping relative to transport across the land border, only Daimler-Chrysler has thus far broken "ranks with its competitors by shipping vehicles manufactured in Mexico across the Gulf of Mexico to the Ports of Tampa and Pensacola, as opposed to transporting them overland through Mexico and Texas" (Petrovsky, 2002). The Gulf of Mexico States Partnership, with funding from the U.S. Congress, commissioned Cambridge Systematics in 2006 to conduct a comprehensive transportation study of the Gulf of Mexico trade corridor. Once completed, the final report may document "increased recognition and use of the water 'border' via the Gulf of Mexico."[3] There appear to be no documented instances of cross-Gulf intrafirm trading or commodity chain integration common to the land border economy.

Although not a direct result of NAFTA, there have been some interesting developments in economic relations with Cuba across the Gulf, for both Mexico and the United States, during the period since the agreement took effect. As pointed out earlier, Mexico has never severed diplomatic or economic relations with Cuba. In discussing Cuba's commercial ties to Mexico, a representative of the Mexican Business Council for International Affairs stated in 1994 that "Cuba is a natural market for Mexico. Cuba is in the Gulf of Mexico . . . two hours by plane from Mexico City . . . one day by boat from the ports of Tampico and Veracruz. Also, Cuba and Mexico have a long history of friendship" (CubaNews, 1994).

With the collapse of the Soviet Union, Mexico became a crucial lifeline for the regime's survival, and Cuba became an eager customer for Mexican companies and investors because the U.S. embargo remained in effect. In the early 1990s major Mexican companies entered Cuba through debt-equity swaps. But later—under the threat of the United States invoking provisions of the 1996 Helms-Burton Act, which was designed to discourage foreign investment in Cuba, and against which the Mexican government took a strong stand—Mexican companies reduced or eliminated their investments on the island (Sagebien and Tsourtouras, 1999). The United States also made its entering into a free trade agreement and providing assistance to overcome the 1994–95 peso crisis conditional upon Mexico limiting its Cuban exposure.

While Mexican investment in Cuba may have declined, it still exists, and Mexico continues to be an important trading partner for Cuba, with some of its competitive advantage coming from the reduced shipping time afforded by its Gulf ports. In addition to Mexico's trade with Cuba, one estimate was that "more than 70 percent of all of Cuba's economic trade with the world goes through Mexico, from saltine crackers to cell phones, Coca-Cola to Dell PCs" (Nevader, 2004). The heavy representation of U.S. products in the trade underlines again how Mexico profits at the expense of U.S. firms that cannot do business directly with Cuba because of the embargo. Mexico is also an important source of visitors for the Cuban tourist trade, which has experienced dramatic revival over the past fifteen years.[4]

While Mexican Gulf trade with Cuba is understandable, given Mexico's refusal to break relations, the growth of U.S. trade with the island in recent years is surprising in view of the efforts of the Bush administration to tighten the embargo. Gulf states and ports have profited from this increased trade.

In 2000, Congress passed and President Clinton signed the Trade Sanctions Reform and Export Enhancement Act, which contained a provision allowing the sale of U.S. food and agricultural goods to Cuba on a cash payment basis. The Castro government initially rejected buying U.S. food until the embargo was completely lifted, but the destruction caused by Hurricane Michelle in 2001 gave the Cuban leader an excuse to reverse his position, and by 2002 the United States had become the leading supplier of imported food products to Cuba (Spadoni, 2004). U.S. food exports totaled $474 million in 2004 and $540 million in 2005 in spite of new restrictions requiring pre-departure payment for shipments. By April 2006 Cuba had spent some $1.5 billion on U.S. food products since 2000, when the loophole was opened (Robles, 2006). Approximately 75% of food shipped to Cuba goes through the Gulf ports, which gives U.S. exporters an estimated 20% cost advantage over European and Canadian suppliers (Spadoni, 2004). While publicly justifying U.S. food purchases in economic terms, the Cuban government is quick to exploit the procedure for political ends, for instance by spreading purchases around the United States in an effort to build pressure in Congress for dismantling the embargo. In recognition of the fact that her state's ports alone accounted for 56% of all sales to Cuba, in 2005 Louisiana Governor Kathleen Blanco became the first southern governor to visit Cuba and meet with Fidel Castro since the 1959 revolution (CubaNews, 2005). Other official delegations from agricultural states regularly visit Cuba in search of business for their farmers. The chairman of the Port of Corpus Christi called for an end to the U.S. embargo in his speech to the 2006 State of the Gulf Summit in Corpus Christi (Chirinos, 2006).

Institution Building on the Gulf   There is no question that NAFTA has accelerated the economic integration of the United States and Mexico, primarily through the freer flow of trade and investment. It does not appear that the Gulf of Mexico as a region is experiencing deeper economic integration than is occurring at the binational level. Additional steps toward forging a transnational community on the Gulf would consist of agreements establishing organizations—both governmental and nongovernmental—linking the Gulf states of the United States and Mexico closer together. In general, NAFTA itself represents a modest step in institution building, since the agreement was "a minimum one that reflected the Canadian and Mexican fear of being dominated by the United States and the U.S. antipathy toward bureaucracy and supra-national organizations" (Pastor, 2005b). Attempts at building special transnational institutional linkages between the U.S. and Mexican Gulf states have been ineffectual. The Gulf institutional framework consisted of the following organizations as of 2006:

> Gulf States Accord
> Gulf Governors Environmental Alliance
> Gulf of Mexico States Partnership, Inc.
> Gulf of Mexico Congressional Caucus
> State of the Gulf of Mexico Summit/Gulf of Mexico Alliance

The only U.S.-Mexican quasi-governmental organization with a clear Gulf orientation is the Gulf States Accord, or GOMSA. In May 1995 the governors of Florida and Mississippi, along with senior officials from Texas, Alabama, and Louisiana, met with their counterparts from six Mexican Gulf states (Tamaulipas, Veracruz, Tabasco, Campeche, Yucatán, and Quintana Roo), and Mexican President Ernesto Zedillo, to draft a charter for GOMSA (Rosenberg, 1995). It called for closer cooperation between the U.S. and Mexican Gulf states in investment, communication and transportation, health, education and culture, and agriculture and tourism, with progress to be monitored through annual meetings of the Gulf governors that would alternate between the two countries. The overall goal of the accord was to "foster, promote, and implement cooperative relationships between and for the mutual benefit of the member states and their private sector communities in support of . . . NAFTA." Specific objectives included developing a "formal mechanism for member states to coordinate joint or bilateral economic development activities," promotion of opportunities for the private sector, and sharing of "information on regulations, laws, trade data and customs" (from draft of Gulf States Accord, January 19–20, 1995, Miami). In his 2005 update, the secretary-general of GOMSA said the organization's goal was to "maximize the Gulf of Mexico states' advantages and strengths as 'border states' under the North American Free Trade Agreement (NAFTA), unifying them as a productive, prosperous, and growing 'neighborhood'" (Springer, 2005).

From the outset GOMSA has struggled to define its identity and mission and to secure funding. Although initiated by a group of Gulf state governors, it is not an intergovernmental organization but a compact with no legal standing. More damaging has been the lack of participation by U.S. Gulf state governors. This means that Mexican Gulf state governors have dominated GOMSA, which seriously undermines its effectiveness as a transnational organization addressing significant issues. A March 2005 newspaper story on Governor Blanco of Louisiana assuming the presidency of GOMSA reported that the organization had "labored in obscurity for years but that Louisiana officials hope to revitalize it as an engine of economic development." The article went on to say that the accord "produced mostly picayune results" (Varney, 2005).

GOMSA currently maintains the original six working groups, but it has refocused its attention on environmental challenges to the Gulf of Mexico. Under GOMSA auspices, the binational Gulf Governors Environmental Alliance was set up, and GOMSA collaborates with a U.S. Environmental Protection Agency red tide project on the Gulf. The 2005 Gulf hurricane season reinforced the need for closer collaboration among Gulf states on natural disasters, but it also made such collaboration difficult in the short run because the Gulf states had to focus on short-term recovery. Overall, it is clear that GOMSA has not lived up to its promise of building institutional ties between Gulf state governors across the U.S.-Mexican border.

The Gulf of Mexico States Partnership, Inc. is a "business advocacy and research partner to GOMSA" (Springer, 2005). As a private organization with legal standing in the United States, it is able to raise funds for research and advocacy on issues affecting the Gulf. Its Gulf of Mexico Trade Corridor Transportation Study is described as "the first of its kind focused on both sides of the Gulf of Mexico [which would] create a long-range blueprint for enhancing homeland security and deepening the economic benefits of the North American Free Trade Agreement in the strategic Gulf of Mexico basin trade corridor" (http://www .gulfofmexicostatespartnership.com/study.html). The Partnership hosts the General Secretariat of GOMSA, and it collaborates with the Gulf of Mexico Caucus in the U.S. Congress, which is an officially registered congressional caucus but which lost its leader when Representative Katherine Harris gave up her House seat to run for the Senate in Florida in 2006.

An initiative directed in part at addressing the widely acknowledged institutional weakness of Gulf state collaboration was the March 2006 State of the Gulf of Mexico Summit hosted by the Harte Research Institute for Gulf of Mexico Studies at Texas A&M University–Corpus Christi. The overall goal of the summit was to "focus attention, discussion, and collaborative action on achieving sustainable economics and environmental quality within the Gulf of Mexico Region." Specific objectives included: (1) engaging "Gulf State Governors from the United States and Mexico in developing proactive programs addressing the challenges of sustainable economies and healthy marine environments"; (2) introducing "the Gulf of Mexico Alliance, a communications framework for integrated cooperative regional governance"; and (3) establishing "working committees to carry forth development and implementation of an action agenda as directed by the Gulf State Governors addressing the challenges of sustainable economies and healthy marine environments" (http://www.stateofthegulf.org/ conference.htm). The proposed Gulf of Mexico Alliance closely resembles the Gulf States Accord, both in mission and organization, which raises two questions: Why is a second Gulf of Mexico organization needed, and why should we expect the Alliance to be more effective than GOMSA?

Conclusion

This chapter's analysis leads to the conclusion that the Gulf of Mexico is neither a transnational community nor on its way to becoming one, even under the auspices of NAFTA. A review of the history of the Gulf region documents that there was little foundation or historical precedence for building regional identification and integration on the Gulf. From the onset of European colonization, the Gulf of Mexico was contested, politically fragmented territory. Independence for the three Gulf nations did not alter this pattern. U.S. efforts to dominate the region through active interference in the internal affairs of its two Gulf neighbors in the first third of the twentieth century generated asymmetry and tension in relations. While U.S.-Mexican relations were largely defined by the land border, U.S.-Cuban relations were conducted across the Gulf. But when Fidel Castro aligned Cuba with the Cold War enemy of the United States, Washington and Havana broke relations, and the strongest, most salient Gulf linkage was severed.

Setting aside Cuba for the moment, consider the state of U.S.-Mexican relations on the eve of NAFTA. In spite of chilly political relations between these two "distant neighbors," economic and demographic forces had drawn them closer together in the post–World War II era. As it was going into operation in the mid-1990s, NAFTA seemed like the next step in the inevitable consolidation of a North American community that would wash away political differences.

There is no doubt that the increased trade and investment flows of the NAFTA era have deepened the economic interdependence of the United States and Mexico. Blank and Golob (2006) argue that "the political economy of North America is no longer composed of three national economies, but rather of links among production cluster and distribution hubs across the continent— links resting on new cross-border alignments among businesses, communities, and local and state-provincial governments." But NAFTA-related economic integration of North America has not enhanced the role of the Gulf in bilateral relations, nor has it been accompanied by the building of transnational political institutions on which a Gulf community could flourish. Part of the explanation for the lack of institution building is the deliberate decision to endow NAFTA with minimal organizational structure in deference to Mexico's fear of the United States dominating the institutions and the U.S. desire to avoid over-bureaucratizing NAFTA. Even without formal institutions like those of the European Union, there were expectations that success in lowering trade barriers would eventually spill over, encouraging the NAFTA states to deepen integration along the path taken in Europe. Indeed, in the early days of the Bush-Fox relationship, Mexicans came to see NAFTA as the first step in constructing a North American community that would address broader issues like immigration and development assistance (Bondi, 2004). In the Guanajato Proposal of February 2001, the two presidents pledged: "After consultation with our Canadian partners, we will strive to consolidate a North American economic community whose benefits reach the lesser-developed areas of the region and extend to the most vulnerable social groups in our countries" (quoted in Pastor, 2005b).

Hopes for movement toward deeper North American integration came to an abrupt end on September 11, 2001, and U.S.-Mexican relations went into a tailspin at the moment that the two neighbors needed each other the most. Not only did the United States turn its attention to the Middle East, but Washington gave President Fox the cold shoulder when Mexico, as a non-permanent member of the U.N. Security Council, refused to support the U.S. position on the Iraq invasion.[5] The near victory of an outspoken critic of Mexico's NAFTA relationship with its northern neighbor in the July 2006 presidential election underlines a delicate state of relations, which continues to feature disagreements over immigration, border security, and drug trafficking. In this context, a North American community that would enhance integration on the Gulf of Mexico is more remote than ever.

The anomalous status of Cuba is an even greater obstacle to Gulf unity. Without the full reintegration of Cuba into the Gulf community, which is not possible until the United States and Cuba restore relations, a transnational community remains out of reach regardless of the state of U.S.-Mexican relations. It is clear, however, that once the embargo is lifted, for whatever reason, the resumption of full U.S.-Cuban diplomatic and trade relations will dramatically affect the Gulf region in multiple ways. Even if a post-embargo Cuba does not join NAFTA, it will once again become a full player in the Gulf of Mexico, and U.S.-Cuban relations could eclipse U.S.-Mexican relations in the Gulf region. Until then, the Cuba question continues to cast a cloud over U.S.-Mexican relations periodically. In February 2006 the U.S. government ordered the U.S. owners of a hotel in Mexico City to expel a Cuban delegation meeting with U.S. oil company officials about offshore exploration opportunities in Cuba's Gulf territorial waters. The Mexican government, which rejects the extraterritorial reach of the embargo, responded by fining the hotel (*Latin America Advisor*, 2006).

In the absence of full Cuban participation in Gulf affairs, and with the persistence of tension in U.S.-Mexican relations, the most promising area for deepening integration of the Gulf of Mexico is collaboration among the Gulf states and provinces to protect the environment. The defining feature of the region is the fragile Gulf ecosystem, which is under the threats of exploding residential development, tourism, overfishing, and accelerated energy exploration. Reacting to rising world oil prices and promising discoveries, all three Gulf nations announced in 2006 that they were stepping up deepwater drilling. The renewed search for oil in the Gulf comes following two hurricane seasons that inflicted tremendous damage in the Gulf and pointed out the environmental risks associated with offshore petroleum exploration. More than 20,000 underwater pipelines and 3,000 offshore platforms were in the path of hurricanes Katrina and Rita in 2005, but "even as they repair the damage, most companies continue to explore the depth of the gulf for new reserves" (Mouawad, 2006). The United States, Mexico, and Cuba—and especially their Gulf territories—all have a stake in the sustainable management of Gulf resources, something that can only be accomplished through coordinated cooperation across national borders. Given the activities already under way, it would not be difficult to forge a regional environmental regime constructed around principles, norms, rules, and decision-making procedures promoting the sustainable development of the Gulf (for a discussion of regime theory, see Krasner, 1983, and Rittberger and Mayer, 1995). Success in deepening collaboration on environmental issues could very well spill over into other issues and thereby advance integration of the Gulf as a transnational community from the bottom up.

Notes

1. Since 1977 Washington and Havana have conducted official business through "interest sections" in the embassies of third countries. Even though the United States does not have full diplomatic relations with Cuba, its delegation is the largest in Havana, occupying a prominent location on the *Malecón,* which makes an inviting target for anti-American demonstrations.

2. In the foreword (page ix) of his 1984 *Distant Neighbors* Alan Riding writes, "Probably nowhere in the world do two countries as different as Mexico and the United States live side by side."

3. For background on the study go to http://www.gulfofmexicostatespartnership .com/study.html.

4. Spadoni (2004) reports that tourist arrivals from Mexico rose from 70,983 to 88,787 from 1999 to 2003, while the total number of foreign tourists grew from 1.6 million to 1.9 million in the same period.

5. "For months Bush did not speak with Fox. Should the Mexican president need any further proof that the big chill was there to stay, in May [2003], the White House cancelled the traditional festivities honoring the Mexican (and Mexican-American) holiday Cinco de Mayo. Petty as it may be, this decision's symbolism was not lost south of the border" (Bondi, 2004).

References

Accord of the States of the Gulf of Mexico. 1995. http://www.gomsa.org/accord/accord .html.

Ávila Sánchez, H. (ed.). 1993. *Lecturas de análisis regional en México y América Latina.* Mexico: Universidad Autónoma Chapingo.

Blank, S., and S. R. Golob. 2006. It Is Time to Talk about North America. *ViewPoint Americas* 4(2). 3 pp.

Bondi, L. 2004. *Beyond the Border and Across the Atlantic: Mexico's Foreign and Security Policy Post-September 11th.* Washington, D.C.: Center for Transatlantic Relations, Johns Hopkins University. 18 pp.

Bretos, M. A. 1991. *Cuba and Florida: Exploration of an Historic Connection, 1539–1991.* Miami: Historical Association of Southern Florida.

Burkholder, M. A., and L. L. Johnson. 1994. *Colonial Latin America,* 2nd ed. New York: Oxford University Press. 384 pp.

Chirinos, F. 2006. Bonilla Calls for End to Cuba Trade Embargo. *Corpus Christi Caller Times.* March 20, p. 1.

CubaNews. 1993–2005. www.CubaNews.com.

Darnell, R. M., and R. E. Defenbaugh. 1990. Gulf of Mexico: Environmental Overview and History of Environmental Research. *American Zoologist* 30: 3–6.

Dominguez, J. I., and R. Fernandez de Castro. 2001. *The United States and Mexico: Between Partnership and Conflict.* New York: Routledge. 184 pp.

Gannon, M. 1993. *Florida: A Short History.* Gainesville: University Press of Florida.

Gore, R. H. 1992. *The Gulf of Mexico: A Treasury of Resources in the American Mediterranean.* Sarasota: Pineapple Press.

Grayson, G. W. 1980. *The Politics of Mexican Oil.* Pittsburgh: University of Pittsburgh Press.

Jackson, J., R. S. Weddle, and W. DeVille. 1990. *Mapping Texas and the Gulf Coast: The Contributions of Saint-Denis, Olivan and LeMaire.* College Station: Texas A&M University Press.

Krasner, S. D. (ed.). 1983. *International Regimes.* Ithaca: Cornell University Press. 384 pp.

Krauss, C. 2006. Big Oil Find Is Reported Deep in Gulf. *New York Times.* September 6.

Langley, L. D. 1989. *America and the Americas: The United States in the Western Hemisphere.* Athens: University of Georgia Press.

Latin American Advisor. March 27, 2006, p. 2.

McCoy, T. L. 1996. The Gulf of Mexico: A Regional Overview. Occasional Research

Paper no. 101. Miami: Latin American and Caribbean Center, Florida International University.

Meyer, M. C., and W. L. Sherman. 1991. *The Course of Mexican History*, 4th ed. New York: Oxford University Press. 718 pp.

MMS (Minerals Management Service). *Gulf of Mexico Update: July 1992–June 1994.* Minerals Management Service.

Mouawad, J. 2006. Divers Work the Gulf Floor to Undo What Hurricanes Did. *New York Times*, March 1.

Nevader, L. 2004. In Rift with Mexico, Cuba Is the Loser. *Pacific News Service*, May 24.

Novel, V. 1994. Culturas viajeras. El intercambio cultural entre Yucatán y Cuba: Una historia no escrita. *Del Caribe* no. 24/94: 55–60.

Pastor, R. A. 2005a. North America and the Americas: Integration among Unequal Partners. Chapter 13 *in* M. Farrell, B. Hettne, and L. Vangenhove (eds.), *Global Politics of Regionalism: Theory and Practice.* Ann Arbor, Mich.: Pluto Press. 336 pp.

———. 2005b. North America: Three Nations, a Partnership or a Community. *Jean Monnet/Robert Schuman Paper Series,* vol. 5, no. 13 (June). University of Miami.

Pérez, L. A. 1988. *Cuba: Between Reform and Revolution.* New York: Oxford University Press. 464 pp.

Petrovsky, M. 2002. Tampa Zeroes in on Mexico Trade. *Gulf Shipper,* February 11.

Raat, W. D. 1992. *Mexico and the United States: Ambivalent Vistas.* Athens: University of Georgia Press.

Randall, L. 1989. *The Political Economy of Mexican Oil.* New York: Praeger. 238 pp.

Riding, A. 1984. *Distant Neighbors: A Portrait of the Mexicans.* New York: Vintage Books. 432 pp.

Rittberger, V., and P. Mayer. 1995. *Regime Theory and International Relations.* Oxford: Clarendon Press. 494 pp.

Robles, F. 2006. Sales of U.S. Food up 20%. *Miami Herald,* April 14, p. 13A.

Rosenberg, M. 1995. The Forgotten Gulf. *Florida Trend,* July, pp. 12–14.

Sagebien, J., and D. Tsourtouras. 1999. Solidarity and Entrepreneurship: The Political-Economy of Mexico-Cuba Commercial Relations at the End of the Twentieth Century. Revised version of paper presented at the Association for the Study of the Cuban Economy Conference, Miami, August 1998. 17 pp.

Salpukas, A. 1995. The New Gulf War: Man Your Computers. *New York Times,* June 18, pp. F1+.

Sheppard, D. M. 1994. The Influence of Meteorological and Oceanographic Conditions on Recreational Boat Routes between the United States and Cuba. In James E. Cato (ed.), *The Potential Impact on Florida-Based Marina and Boating Industries of a Post-Embargo Cuba.* Florida Sea Grant College Program, Technical Paper 76. 147 pp.

Smith, P. H. 2000. *Talons of the Eagle: Dynamics of U.S.–Latin American Relations,* 2nd ed. New York: Oxford University Press.

Smith, W. S. 1987. *The Closest of Enemies: A Personal and Diplomatic History of the Castro Years.* New York: W. W. Norton and Company.

Spadoni, P. 2004. U.S. Financial Flows in the Cuban Economy. *Transnational Law and Contemporary Problems* 81: 81–117.

Springer, G. L. 2005. Integrating the Gulf of Mexico Border. An Update Given on June 28, 2005 at the World Trade Center of New Orleans by President of the Gulf of Mexico Partnership, Inc. and Secretary-General, Gulf of Mexico States Accord. Available at http://www.gomsa.org/.

Varney, J. 2005. Blanco Takes Helm for Gulf-State Group. *Times-Picayune,* March 8, p. 4.

2 The Productive Value of the Gulf of Mexico

DAVID W. YOSKOWITZ

Introduction

The Gulf of Mexico has been referred to repeatedly as the most productive body of water in the United States. But exactly what does this mean? Is there a value that can be placed on this *productivity?* If so, how large is it and what use would this value be? The goal of this chapter is to answer these questions and, in so doing, to highlight the important role that the Gulf of Mexico plays in the economic lives of the United States, Mexico, and Cuba. For the purposes of this analysis the sectors of focus will be oil and gas production, port and shipping activity, tourism, and fisheries, during the year 2003, the most recent year for which complete data are available. There are, of course, a number of other activities, but focusing on these four critical industries offers a good overview of Gulf of Mexico economics. While it would be appropriate to include information about Cuba, because data from that country are not readily available, the discussions focus strictly on the United States and Mexico.

Previous Work

Despite the oceans' importance to us, and the fact that they cover two-thirds of the planet, relatively little work has addressed the oceans' economic importance. To do that, we must begin by defining what we mean by the economic value of our oceans.

The *ocean economy* is that portion of the economy that relies on the ocean as an input to the production process, or that takes place on or under the ocean. The *coastal economy* is that portion of economic activity that takes place on or near the coast (Colgan, 2004).

These definitions have been used in the analysis of regional economies and are very useful for the policymaker. The government of Canada, for instance, conducted a study of the economic value of the ocean sector for Nova Scotia (Gardner Pinfold Consulting Economists, 2005). The gross domestic product (GDP) impact of the ocean sector in that economy was estimated at $1.65 billion ($2.62 billion Canadian) in 2001 and $2.56 billion ($4.08 billion Canadian) when spin-off effects were considered. This accounted for about 15% of Nova Scotia's GDP.

A similar assessment of California's coastal and ocean economy (Kildow and Colgan, 2005) estimated that the value of that state's ocean economy was approximately $42.9 billion for the year 2000. This accounted for 19% of the national ocean economy, when compared to the aggregated values for the United States economy.

Although the concepts of the ocean and coastal economies are close to the idea of a productive value for the ocean, they still fall short in application to the question of the productive value of the Gulf of Mexico. The *productive value,* as defined for the purposes of this study, is the market value of the resources extracted from the Gulf, or in the case of tourism and port operations, it is the value of the services generated as a result of proximity to the Gulf. Unlike the broader definitions of the ocean and coastal economies, the term *productive value* does not use a multiplier of any sort for income or employment. This deliberately narrower concept, therefore, offers a clearer picture of the value directly derived from the Gulf of Mexico itself.

The Approach

Data for some industry sectors covered in this chapter are not readily available, so significant research is required to generate the values. It is important to note that the definitions for certain terms involved in calculations are not always consistent between the United States and Mexico. Nonetheless, the productive values presented here are a positive first step that represents a lower bound or conservative minimum.

To place the productive values generated in proper context, it is essential to understand how relevant data were obtained. A description of the methodological approach used for each sector is therefore provided in the sections that follow.

Oil and Gas Production   U.S. oil and gas production occurs at two main location types: federal leases and state leases. Production levels on federal leases were obtained from the Minerals Management Service (MMS, 2006), while production levels on state leases were acquired from the appropriate state regulating bodies. Production levels for oil were then multiplied by average prices in 2003 of $28.50 per barrel for oil and, for natural gas, $4.88 per thousand cubic feet.

In Mexico, oil and gas are controlled by the state monopoly PEMEX. Its earnings from oil and gas production are reported in the company's 2003 annual report (PEMEX, 2003).

Fisheries   Conveniently, both the U.S. and Mexican governments provide easily accessible data and dockside values for commercial catches, and the totals for these data are used as the productive value for this industry. In the United States, data comes from the National Marine Fisheries Service (2004), while in Mexico they are provided by the Comisión Nacional de Acuacultura y Pesca (2003) in their annual report.

Ports and Shipping   Assessing the value of Gulf shipping is a daunting task. Shippers use the Gulf waters as a conveyance mechanism like truckers use roads, and it is difficult to assess the value of such use. One way to measure this value is to calculate how much revenue shippers derive from the Gulf, but shipping to and from the Gulf also involves shipping to and from the rest of the world. The task therefore becomes collecting revenue figures for each shipper and parsing out the fraction of the total that can be attributed strictly to the distance traveled in Gulf waters. For example, if coffee is shipped from Colombia to New Orleans and the fee for shipping is X dollars, then the productive value is the fraction of the total distance that is in Gulf waters multiplied by X dollars.

Unfortunately, acquiring the data needed to make such calculations is

Table 2.1. *Port Revenue (in U.S. $)*

Mexican Ports	Operating Revenue	
Altamira	11,707,900	
Coatzacoalcos	11,293,233	
Dos Bocas	12,089,600	
Progreso	822,358	
Tampico	10,957,080	
Tuxpan	4,237,602	(2004)
Veracruz	14,285,858	(2004)

United States Ports	Operating Revenue	
Alabama		
Alabama State Port Authority	60,092,334	
Florida		
Port Manatee	8,180,606	
Port of Pensacola	1,474,175	
Port of St. Petersburg	228,000	
Tampa Port Authority	21,373,000	
Lousiana		
Greater Baton Rouge Port Commission	2,456,400	
Greater Lafourche Port Commission	9,300,000	
Lake Charles Harbor and Terminal District	15,590,573	
Port of New Orleans	33,500,000	
Port of Shreveport	503,819	
Port of South Louisiana	6,641,000	
St. Bernard Port	3,160,000	
Mississippi		
Mississippi State Port Authority	9,815,849	
Port of Pascagoula	2,498,000	(2001)
Texas		
Port of Beaumont	11,684,783	
Port of Corpus Christi	19,199,014	
Port of Freeport	6,184,898	
Port of Galveston	7,186,000	
Port of Houston	101,085,000	
Port of Orange	809,512	
Port of Port Arthur	4,980,000	
Port of Port Lavaca	5,409,106	

All data are from 2003 unless noted. Missing from this analysis are Port of Brownsville, Panama City Port Authority, Plaquemines Port, and Port of Iberia.

extremely difficult. Therefore, for the purposes of this study, the operating revenue that ports generate from vessel and cargo services is used as a proxy for the productive value of Gulf shipping. Operating revenues exclude revenue from the renting of warehouses or other moneymaking activities not associated with shipping and are available on financial statements for ports in both the United States and Mexico (table 2.1).

At the time of this study, Hurricanes Katrina and Rita had just made their way through the Gulf. For obvious reasons, some ports in Louisiana were not included in the analysis because data were difficult to come by, given that the attention in the ports was focused on the aftermath of the storms. Therefore, the total value in this category slightly underrepresents the complete value. Complete information was available for the Mexican ports.

Tourism   For four of the five U.S. Gulf states, estimating tourism revenues for counties directly adjacent to the Gulf, and therefore deriving a product from the Gulf, is relatively straightforward because state tourism offices are well equipped and eager to collect and report these figures (table 2.2). For Texas, figures were provided by the Governor's Economic Development and Tourism office in *The Economic Impact of Travel on Texas* (Texas Tourism Office of the Governor, 2005). In Louisiana, it was the Office of Tourism (2004) and *Economic Impact of Travel on Louisiana Parishes, 2003.* The Mississippi Development Authority/Tourism Division (2004) published *Fiscal Year 2003 Economic Impact for Tourism in Mississippi,* while the Alabama Bureau of Tourism and Travel (2005) issued *Economic Impact of the Alabama Travel Industry 2004.*

Unfortunately, Florida does not collect impact or expenditure data at the county level in the same manner as the other Gulf states. Florida expenditures related to tourism were therefore derived from the tax receipts received from the designated *tourist development tax.* This is essentially a *bed tax* levied on charges from hotels, motels, rooming houses, and apartments. All but one of the coastal counties in Florida had such a tax in 2003. In order to derive the total expenditures, the *tourist* tax receipts for each of the coastal counties were divided by the *tourist development tax* rate, which ranges from 2% to 4% (table 2.2). Because it is impossible to separate the tax revenue generated by tourists from that generated by residents or business travelers, this method overestimates the amount of money generated by tourists in lodging facilities. However, the method underestimates the total amount of money generated by tourism because it does not take into account expenditures in other categories. Therefore, if anything, the values on tourism expenditures for Florida are very conservative.

Mexican tourism data at the state or municipio level are limited, with a few exceptions. The Secretaria de Turismo (SECTUR) and the Instituto Nacional de Estadistica Geografía e Informática (INEGI) do a complete job of tracking national tourism trends in aggregate form, but they do not release data for individual areas as would be needed to separate out tourism tied directly to the Gulf of Mexico. The number of visitors in 2004 for Mexico's Gulf states, with the exception of Tamaulipas, is available. To estimate a tourism productive value for Mexico, this number was multiplied by the average expenditure per tourist per visit in Mexico, estimated by SECTUR (SECTUR, 2004) at $673.70.

Table 2.2. *Tourism Revenue*

State	Tourism Revenue
Alabama[a]	$2,378,195,976
Florida[b]	$15,450,000,000
Louisiana[c]	$5,263,550,000
Mississippi[d]	$1,756,800,418
Texas[e]	$10,039,916,000

[a] Alabama Bureau of Tourism and Travel, 2005
[b] Author's own calculations
[c] Louisiana Office of Tourism, 2004
[d] Mississippi Development Authority/Tourism Division, 2004
[e] Texas Tourism Office of the Governor, 2005

Table 2.3. *Productive Value (in billions of U.S. dollars)*

	Mexico	United States	Total
Oil and Gas	37.9	39.8	77.7
Fisheries	0.381	0.683	1.064
Port/Shipping	0.0654	0.331	0.396
Tourism	9.965	34.888	44.85
Total Productive Value of the Gulf of Mexico			**124.01**

Discussion

By supplying a snapshot of the economic values for four sectors completely dependent upon the Gulf of Mexico, this chapter constitutes a starting point for understanding the value that the Gulf provides for the United States and Mexico (table 2.3).

One important question to consider is how sensitive these valuations are to changing global economic conditions. Further examination of the oil and gas sector reveals significant sensitivity. As noted, the values presented were based on prices of $28.50 per barrel of crude oil and $4.88 per thousand cubic feet for natural gas. Triple the price of oil to reflect its November 2007 climb to $90 per barrel, and the productive value of the Gulf increases by approximately $110 billion to $234 billion. This is still a conservative productive value for the Gulf of Mexico.

The Gulf also has value beyond quantifiable industries, of course. There are many irreplaceable attributes, such as pristine bays and estuaries, which act as nurseries for a variety of wildlife and commercially valuable fish and shellfish species and as winter habitat for migrating birds. Thus the values provided in this chapter are intended as a starting point for discussion, and much more work is needed. Inclusion of non-market values would be a logical next step in a more exhaustive study calculating the productive value of the Gulf of Mexico (see chapters 4 and 5).

To put the figures that have been calculated in better context, consider that the adjusted productive value of $234 billion

- is greater than the GDPs of Chile, Peru, Finland, and Venezuela (all 2006)
- would put the Gulf of Mexico at 29th out of 230 countries in terms of GDP
- is 27% of Mexico's GDP
- is 1.7% of U.S. GDP
- is greater than the gross state products of Louisiana, Oregon, and Kentucky, to name a few.

As indicated, the Gulf's productive value is not static. It is constantly changing due to several factors, most directly the market values of the resources themselves. In addition, the four sectors covered are impacted by, and impact, other activities. The fisheries industry in particular faces several challenges. In the northern Gulf a zone of low oxygen, or hypoxia, is tied to pollution from the Mississippi River. Known as the *dead zone*, this phenomenon spreads each summer and fades away in the fall. It forces shrimp and certain fish into areas that may be harder and more expensive for fishing boats to reach, and it also has devastating effects on the ecosystem. In 2002 the zone reached a peak size of 22,000 km² (8,500 mi²), an area larger than the state of Massachusetts, and it averages about 17,000 km² (6,500 mi²). Other problems in the Gulf arise from overfishing of species—including red snapper (*Lutjanus campechanus*), red grouper (*Epinephelus morio*), greater amberjack (*Seriola dumerili*), and goliath grouper (*Epinephelus itajara*)—which threatens the species themselves and the long-term viability of the commercial and recreational fishing industries.

Oil and gas firms continually have to go farther and deeper in order to tap new reserves, and it is unclear what the immediate impact of pending moves and expansion will be. The U.S. sector has been buoyed by recent news of a major find in the Walker Ridge area of the Gulf of Mexico that could reportedly increase U.S. reserves by 50%. Mexico is facing the daunting challenge of finding new reserves as PEMEX continues to squeeze production out of its existing Reforma-Tabasco fields and Campeche Sound. The country is scrambling to find the resources and technology needed to continue output at the same level or even to increase it over time. Prices will continue to play a critical role not only in determining the productive value of the Gulf but also in providing an incentive for continued exploration into deeper and more expensive reserves.

Coastal development will continue to be blessed by economic development offices, cursed by environmentalists, and stressful for insurance companies. The events of the 2005 hurricane season for the northern, western, and southern Gulf provided ample evidence of the human and infrastructure costs of being close to the water.

In the United States, the Census Bureau estimates that the population of the five Gulf Coast states will increase from 1995's total of 44.2 million to an estimated 61.4 million in 2025. Similar growth is expected in the Gulf states in Mexico and Cuba. Even though the development of the Gulf's coast is greatly boosting the economic growth of the region, the high density of people and industry is also a potential threat to the ecological condition of the Gulf's diverse and productive coastal environment. This includes estuaries, coastal wetlands, coral reefs, mangrove forests, and upwelling areas where nutrient-rich deep-ocean water rises to the surface, providing food for algae and other organisms near the surface. These resources are critical for supporting healthy populations of a wide range of marine life as well as birds and other wildlife (Nipper et al., 2006).

All indicators suggest that tourism will continue to drive a large part of coastal development, and as a result, population density will be greater in the region, placing added stress on a fragile ecosystem. Already, the Gulf and its beaches are being polluted by hundreds of thousands of gallons of oil and hazardous materials that spill into the water annually, and more than 500 tons of trash washes ashore each year (University of Texas at Austin and Texas State Historical Association, 1999). Another concern is erosion occurring at many beaches, which can threaten property and reduce a beach's attraction for tourists. It remains to be seen whether tourism will continue to thrive if what the tourists are coming for begins to disappear. Likewise, it is not clear whether increased development will put long-term growth at risk.

Conclusion

While the productive value for the Gulf of Mexico was calculated by examining each industry in isolation, there is no doubt about the extensive interconnectedness between these industries, the Gulf ecosystem, coastal populations, and other sectors of the economy. And though the economic environments of the United States and Mexico are presented separately, they are very much tied together and also impacted by the Cuban economy.

Here are some steps that could improve the dialogue among the nations in the Gulf region and, in the process, strengthen policymaking and economic opportunity:

(1) Increasing understanding of the economic interconnectedness and interdependency between the Gulf states of Cuba, Mexico, and the United States.
(2) Analyzing the economic and ecological linkage in the Gulf by combining information and results from both the economics and marine ecology fields into a single model for improved policymaking. The model should include a full range of economic sectors and trophic levels in a linked ecosystem.
(3) Standardizing and reporting economic data relevant to the Gulf of Mexico and then centralizing and maintaining this important information to aid industry, policymakers, and academics.

References

Alabama Bureau of Tourism and Travel. 2005. *Economic Impact of Alabama Travel Industry 2004*. Montgomery: Alabama Bureau of Tourism and Travel. 80 pp.

Colgan, C. S. 2004. *The Changing Ocean and Coastal Economy of the United States: A Briefing Paper for Governors*. Monterey, Calif.: National Ocean Economics Project. 18 pp.

Comisión Nacional de Acuacultura y Pesca. http://www.conapesca.sagarpa.gob.mx/wb/cona/cona_anuario_estadistico_de_pesca.

Gardner Pinfold Consulting Economists. 2005. *Economic Value of the Nova Scotia Ocean Sector*. Prepared for the Government of Canada and Nova Scotia. Halifax, Nova Scotia: Gardner Pinfold Consulting Economists. 80 pp.

Kildow, J., and C. Colgan. 2005. California's Ocean Economy. Monterey, Calif.: National Ocean Economics Project. 156 pp.

Louisiana Office of Tourism. 2004. *Economic Impact of Travel on Louisiana Parishes, 2003*. Baton Rouge: Louisiana Office of Tourism. 47 pp.

Mississippi Development Authority, Tourism Division 2004. *Fiscal Year 2003 Economic Impact for Tourism in Mississippi*. Jackson: Mississippi Development Authority, Tourism Division. 52 pp.

MMS (Minerals Management Service). http://www.gomr.mms.gov/homepg/pubinfo/repcat/product/pdf/Annual%20Production%202000%20-2004.pdf.

National Marine Fisheries Service. http://www.st.nmfs.gov/st1/fus/fus04/02_commercial2004.pdf.

Nipper, M., J. A. Sánchez Chávez, J. W. Tunnell Jr. (eds.). 2006. GulfBase: Resource Database for Gulf of Mexico Research. http://www.gulfbase.org.

PEMEX. 2003. *Annual Report 2003*. Mexico City: PEMEX. 64 pp.

SECTUR. 2004. Resultados acumulados de la actividad turística, Enero–Diciembre, 2004. 18 pp.

Texas Tourism Office of the Governor. 2005. *The Economic Impact of Travel on Texas*. Austin: Texas Tourism Office of the Governor. 164 pp.

University of Texas at Austin and Texas State Historical Association. 1999. *Handbook of Texas Online*. Austin: University of Texas at Austin and Texas State Historical Association. http://www.tsha.utexas.edu/handbook/online/articles/GG/rrg7.html.

3 An Economic Overview of Selected Industries Dependent upon the Gulf of Mexico

CHARLES M. ADAMS, EMILIO HERNANDEZ, AND JIM LEE

Introduction

The Gulf of Mexico is a critical source of natural resources, providing billions of dollars in tangible and intangible benefits to a variety of marine-related industries and other user groups. The economic benefits of Gulf resources flow not only to bordering states but also to the U.S. economy as a whole. Industries directly or indirectly dependent on the Gulf ecosystem include coastal development, coastal recreation and tourism, merchant shipping, offshore oil and gas production, hard mineral mining, recreational boating, and commercial fisheries. Some of these, such as commercial and recreational fishing and tourism, are entirely dependent on a healthy coastal and marine ecosystem for their existence. The rapidly growing coastal population and industrial base are placing increasing demands on the Gulf's critical natural resources. As a result, resource managers are becoming increasingly aware of the need for aggressive measures to enable sustainable management of Gulf resources and to ensure that marine-related user groups and industries have future access to Gulf of Mexico natural resources.

Such proper management of Gulf natural resources is, in part, dependent on an adequate understanding of the economic value derived from these resources. To that end, this chapter is intended to provide a brief overview of economic values and activities of key U.S. industries dependent on those resources by focusing on important trends in past decades for those industries. The chapter is not intended to be an exhaustive assessment of all the marine-related industries in the Gulf region, but instead focuses on

> petroleum extraction
> commercial fishing
> commercial seafood processing
> marine sportfishing
> merchant shipping
> cruise industry activity
> maritime vessel construction
> marine recreational activities

The discussion draws heavily from a previously published article by Adams et al. (2005) but, where possible, provides more recent data and trends. In addition, this overview focuses on the period before the devastating storm events of the 2004 and 2005 hurricane seasons. With the exception of the data for the petroleum industry, the information needed to quantify the short-term effects of the storms was not available at the time of writing. Also, the long-term effects of the Hurricanes Katrina and Rita have yet to be fully realized by most industries and their associated communities within the Gulf region.

The industries addressed make significant contributions to the Mexican and Cuban economies, some of which are described elsewhere in this book. However, this chapter focuses solely on impacts to the U.S. economy.

Petroleum Extraction

Oil and gas reserves are key Gulf of Mexico economic assets that support such operations as oil and natural gas exploration and extraction, oil refineries, petrochemical and natural gas processing, supply and service bases for offshore oil and gas production, offshore platform and pipeline construction, and other industry-related installations (French et al., 2006). Offshore oil and gas production in the Gulf of Mexico occurs mainly near the coastlines of Louisiana and eastern Texas, though there is limited production as far east as Alabama.

The offshore oil and gas industry has endured dramatic changes over the last thirty years due to such factors as international finances, political decisions and actions, and changes in domestic consumer demand for petroleum-based fuels. Recent proposed lease sales off the Florida coast, as well as proposed liquefied natural gas (LNG) facilities in the Gulf, have drawn new attention to both the environmental and economic values of the Gulf coastal areas. If managed improperly, such expanded activities have the potential to degrade Gulf resources. Therefore, balancing environmental concerns with the economic and national security benefits associated with increased offshore oil, gas, and mineral production is of utmost concern to resource managers, public policymakers, and industry executives within the region.

Production Trends   Historically, oil and gas production in the Gulf have undergone cyclical swings. In addition to weather conditions, oil and gas exploration and drilling activities have been affected by many factors, including technological improvements, price fluctuations, general economic conditions, and financial and political developments in the United States or abroad. Gulf offshore oil production in federal waters climbed to a production peak of approximately 370 million barrels per year in the early 1970s, and then achieved another peak of about 350 million barrels per year during the mid-1980s (fig. 3.1). After declining during the late 1980s, solid productivity gains through technological improvements allowed oil production to grow steadily throughout the 1990s, reaching a peak of 567 million barrels in 2002 (MMS, 2006).

In contrast to oil, natural gas production in the Gulf region has been relatively stable since the late 1970s (fig. 3.1). From its beginning in the 1950s, natural gas production offshore exhibited a continuous increase until peaking in 1981 at 4.9 trillion cubic feet. Natural gas production then decreased, with considerable volatility in production between 1982 and 1986. Gas production increased to 4.5 trillion cubic feet in 1987. During the period from 1987 to 2001, gas production was more stable and averaged approximately 4.8 trillion cubic feet per year. However, since 1997 there has been a moderate decrease in production, falling to 4.4 trillion cubic feet by 2003.

Although Gulf region oil and gas production were both diminishing by 2003, Hurricanes Katrina and Rita exacerbated the downward trend. The average number of active rotary drilling rigs operating in offshore Gulf waters at any time during 2003 was 86, 18, and 1 for Louisiana, Texas, and Alabama, respectively (Baker Hughes, 2006). Just prior to the landfall of Hurricane Katrina in 2005, 79% of the Gulf platforms were evacuated. In addition, 95% of oil (1.4 million barrels per day) and 88% of natural gas production (8.8 billion cubic

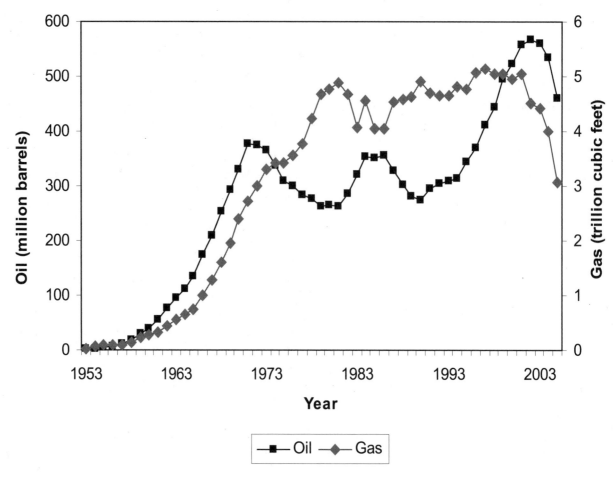

Figure 3.1. Gulf of Mexico oil and gas production volumes, federal waters, 1953 to 2005. Source: MMS, 2006.

feet per day) in the Gulf of Mexico offshore industry was disrupted due to safety reasons, or *shut-in*. (Energy Information Agency, 2006). Three weeks after Hurricane Katrina, approximately 55% of oil and 34% of natural gas production was still shut-in. When Hurricane Rita hit less than a month later, shut-in capacity once again spiked up to Katrina levels. Hurricanes Katrina and Rita caused a production loss estimated at 188 million barrels of refined petroleum products by the end of 2005 (Darby et al., 2006).

Economic Activities Associated with Offshore Petroleum Production Since 1992, approximately 89% of the oil production and 74% of the natural gas production in the Gulf of Mexico has come from the central Gulf region, which includes Louisiana, Mississippi, and Alabama. However, the vast majority of the oil and gas production within the central Gulf region is associated with wells and platforms off the coast of Louisiana (MMS, 2006). As such, the majority of the economic activity associated with offshore oil and gas production in the Gulf of Mexico is tied to the Louisiana industry. Though the entire oil and gas industry provides significant economic benefits for Louisiana (Scott, 2002), the offshore oil and gas operations located off the coast of Louisiana contribute $6 billion annually to the state's economy (Applied Technology Research Corporation, 1999). More than $1.2 billion of this total is salaries and wages paid to offshore workers, and at least 21,000 oil industry jobs are created by the offshore oil and gas industry. Similarly, offshore natural gas exploration activities located off the Alabama coast generate over 7,000 jobs and $4 billion in industry-related expenditures each year (Plater et al., 1999).

Petroleum Production and Recreation   Interestingly, the offshore oil and gas industry also indirectly supports marine sportfishing and recreational diving activities in the Gulf of Mexico. Many of the 5,000 offshore oil and gas structures in the Gulf provide fishermen and divers with destinations for their activities, which supply direct benefits to the coastal economies of U.S. Gulf states. Hiett and Milon (2001) found that fishing and diving activities associated with oil and gas structures generated $325 million in economic output in the coastal counties within the Gulf region. In addition, approximately 5,560 full time jobs are created by these fishing and diving activities.

Commercial Fishing and Seafood Processing/ Wholesaling

The commercial fishing industry is an important component of the Gulf of Mexico ecosystem's total economic value. The commercial fishing industry in the Gulf harvested 1.4 billion pounds in whole weight of fishery products during 1992, valued at $652 million dockside, which is the value received by the vessel or boat upon offloading to a first handler shoreside (NOAA, 2004). Since 1992 Gulf landings have exhibited a somewhat erratic trend (fig. 3.2). Landings peaked at 2.2 billion pounds in 1994 and then remained between 1.5 and 2.0 billion pounds through 2003, when landings totaled 1.6 billion pounds. Total nominal (not adjusted for inflation) dockside value of annual landings has

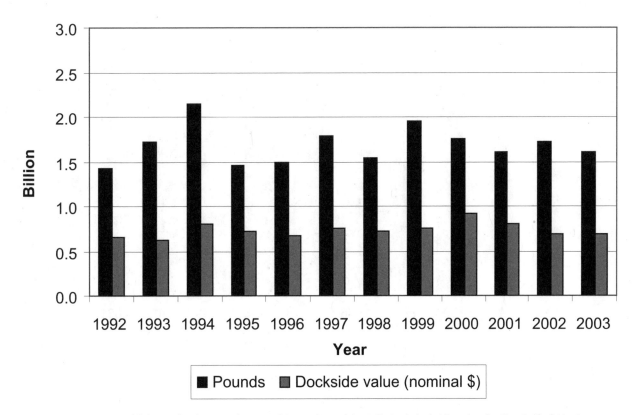

Figure 3.2. Commercial fisheries landings and nominal (not adjusted for inflation) dockside value for the Gulf of Mexico, 1992 to 2003 (in billions). Source: NOAA, 2004.

remained somewhat steady from 1992 to 2003, averaging $735 million over the twelve-year period. Dockside value totaled $652 million in 1992, peaked at $911 million during 2000, and then fell to $683 million during 2003.

Gulf Landings and Dockside Value Trends   Between 1992 and 2003, the commercial fishing industry in the Gulf of Mexico accounted for an average of approximately 21% of the U.S. seafood landings and about 18% of the total U.S. dockside value (fig. 3.3). Annual totals fluctuated during that period, with Gulf landings and value in 1992 at 15% and 18% of the U.S. total, respectively. These shares increased to 17% and 20%, respectively, by 2003 (NOAA, 2004).

Of the total Gulf region landings, Louisiana is by far the largest contributor of any Gulf state (fig. 3.4), with menhaden (*Brevoortia patronus*), a finfish useful for industrial purposes, dominating its catches. Louisiana landed 1.2 billion pounds of fishery products in 2003, compared to 213 million pounds for Mississippi, 79 million pounds for the west coast of Florida, 96 million pounds in Texas, and 25 million pounds for Alabama. From 1992 to 2003, Louisiana and Mississippi landings increased by 8% and 15%, respectively; Florida landings decreased by 27%; and landings for both Texas and Alabama stayed about the same. Overall, landings in the Gulf region declined by about 11% between 1992 and 2003.

Due to constantly fluctuating prices for fishery products, trends for nominal dockside value differed in some cases from those seen in landings. From 1992 to 2003, total nominal dockside value for the Gulf region as a whole increased

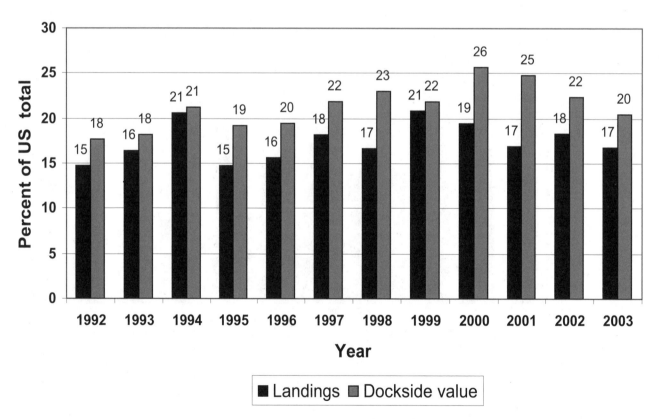

Figure 3.3. Gulf of Mexico commercial fisheries landings and value as a percentage of United States total, 1992 to 2003. Source: NOAA, 2004.

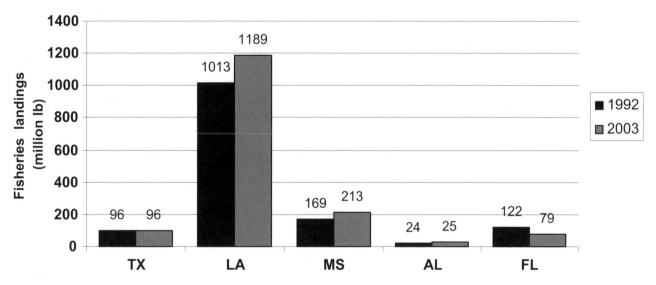

Figure 3.4. Commercial fisheries landings for the Gulf states, 1992 and 2003 (in millions of pounds). Source: NOAA, 2004.

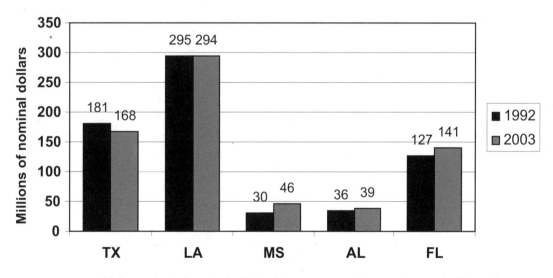

Figure 3.5. Commercial fisheries dockside value for the Gulf states, 1992 and 2003 (in nominal millions of dollars). Source: NOAA, 2004.

by approximately 5%. During this period, Mississippi's dockside value increased by 53%, Alabama's by 8%, and the west coast of Florida's by 11%. Texas saw a 7% decrease, and Louisiana remained virtually the same (fig. 3.5).

Vessels and Boats   A total of 20,470 commercial fishing craft were accounted for within the Gulf region during 2002, representing approximately one-third of the nation's entire commercial fishing fleet. The Gulf of Mexico fleet is extremely diverse and comprises vessels with a wide range of sizes, construction types, gear types, and harvest technologies. The open-water Gulf harvest fleet includes large shrimp trawlers, purse seiners, and longliners outfitted with large freezer holds that allow them to stay at sea for weeks at a time. In addition, the fleet encompasses an even larger number of small, nearshore craft that use smaller crews and ice holds and take much shorter trips.

The Gulf fleet is composed of *vessels,* formally defined as documented craft with greater than 5 net tons of displacement, and *boats,* which have less than 5 net tons displacement (NOAA, 2004). Louisiana has the largest fleet of both, with 2,084 vessels and 8,874 boats as of 2002. Florida also has a large fleet, with 1,934 vessels and 4,438 boats, but the numbers reported include craft based on both its coasts. Texas reported 4,648 commercial fishing licenses, of which several may be associated with a given vessel (Texas Parks and Wildlife, 2005). The total numbers of vessels and boats for Mississippi and Alabama are 1,365 and 1,775, respectively.

Some vessels and boats move around the Gulf region as commercial seasons, weather, and species availability change. In addition, craft registered outside the Gulf periodically move into the region to fish; the values in this section do not account for these movements.

For some fisheries in the Gulf region, the harvesting sector is considered to be overcapitalized in the number of craft and the collective fishing power they possess. This essentially suggests that there is too much investment capital in the production sector of the fishery. However, recent regulatory and market changes have impacted the number of craft operating in the Gulf region. For example, the number of boats has decreased in the Gulf region due to increased restrictions on the use of nearshore entangling nets. More recently, the number of shrimp trawlers has reportedly decreased due to declining dockside prices and increasing fuel costs. Also, changes in regulations that place further restrictions on allowable gear types, harvest seasons, trip limits, and overall harvest quotas are purported to have reduced the number of craft operating in the Gulf. The number of craft may decrease even more if regulations designed to reduce fishing effort become more stringent and if the U.S. seafood market continues the current trend toward increasing domination by relatively low-cost imported seafood products.

Major Ports   There were eighteen commercial ports in the Gulf region where the volume or dockside value of the seafood offloaded exceeded 10 million pounds or $10 million, respectively, during 2003. The top ten ports, ranked in terms of dockside value, account for one-half of the total volume and dockside value for the Gulf (table 3.1). Four of these ports are in Louisiana, three are in Texas, one is in Mississippi, one is in Alabama, and one is in Florida. The leading port in volume and value is Empire-Venice, Louisiana. The volume offloaded at this port is exceeded in the United States only by that reported for Dutch Harbor-Unalaska, Alaska, and only four other ports in the United States have higher dockside values than Empire-Venice (NOAA, 2004). Nonetheless, the value per pound reported for Empire-Venice landings, as well as for Cameron, Louisiana, is very low because the primary species offloaded at these ports is the low-value menhaden. Shrimp landings dominate other Gulf ports. The value offloaded at Key West, Florida, is higher than all the others, due to landings of high-value pink shrimp (*Farfantepenaeus duorarum*), spiny lobster (*Panulirus argus*), and stone crab (*Menippe mercenaria*).

Seafood Processing and Wholesaling   The Gulf region contains a quarter of the U.S. seafood processing plants and wholesaling establishments. The proces-

Table 3.1. *Major Commercial Fisheries Ports in the Gulf States, 2003*

	Pounds Offloaded*	Dockside* Value ($)	$ / pound
Empire-Venice, LA	400.0	50.8	$0.13
Dulac-Chauvin, LA	39.4	42.3	$1.07
Key West, FL	15.8	38.4	$2.43
Brownsville-Port Isabel, TX	17.9	35.9	$2.01
Galveston, TX	18.6	32.7	$1.76
Bayou La Batrie, AL	18.5	30.8	$1.66
Port Arthur, TX	17.5	30.1	$1.72
Golden Meadow-Leeville, LA	25.5	29.1	$1.14
Gulfport-Biloxi, MS	17.4	26.8	$1.54
Cameron, LA	259.0	25.1	$0.10

Source: NOAA, 2005b.
* Units of one million.

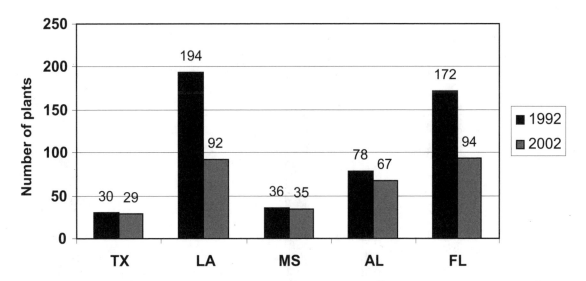

Figure 3.6. Seafood processing plants in the Gulf states, 1992 and 2002. Source: NOAA, 2005a.

sors generate approximately one-fourth of the total value of domestic processed fisheries products, and the majority of the plants are located in Louisiana and Florida (fig. 3.6). Since 1992 the number of processing plants has decreased in every state within the Gulf region, with a decrease of approximately 50% in Louisiana and Florida.

In contrast to the overall decline in the number of processing facilities, the total nominal value of processed fishery products from the Gulf of Mexico has risen from $1.0 billion in 1992 to $1.3 billion in 2003, an increase of 30% (table 3.2). During 2003 Louisiana, Mississippi, and Florida landed 28%, 26%, and 24% of the total Gulf value, respectively, while Texas and Alabama accounted for 13% and 9% of the total processed seafood value, respectively. The volume of processed fishery products also showed an increase during the 1992 to 2003

Table 3.2. *Volume and Value of Gulf Region
Processed Fishery Products, 1992–2003*

Year	Pounds (lb.)	Processed Value ($)
1992	892	1,008
1993	1,045	992
1994	1,189	1,064
1995	896	1,149
1996	896	1,062
1997	1,159	1,433
1998	1,000	1,496
1999	1,184	1,566
2000	1,051	1,636
2001	1,045	1,513
2002	1,049	1,373
2003	989	1,298

Source: NOAA, 2005a.
Units of one million.

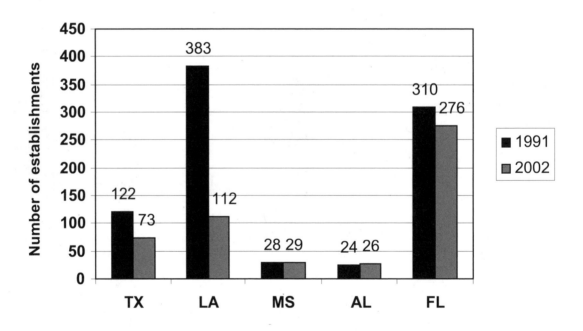

Figure 3.7. Seafood wholesaling establishments in the Gulf states, 1991 and 2002. Source: NOAA, 2004.

period, rising by 11% overall, though volume has decreased steadily since 1999 (NOAA, 2004 and 2005a). To put these statistics in proper context, it is important to note that a significant percentage of the edible seafood processed in the Gulf region is imported from foreign sources or states outside the Gulf region.

The seafood distribution system in the Gulf region comprises a complex network of marketing agents and purveyors. Almost 900 seafood wholesaling establishments were found in the Gulf region during 1991 (fig. 3.7), over 40% of which were in Louisiana. However, since 1991 the number of wholesalers in the region has decreased by almost 40%. Virtually all of this contraction has occurred in Louisiana, where the number of establishments decreased by 70% between 1991 and 2002, though the number of facilities in Texas and Florida has also dropped. These numbers refer only to primary seafood wholesaling facilities and do not account for other types of seafood marketing agents, such as those establishments strictly associated with seafood shipping, brokering, and retailing activities.

Although shrimp account for a large portion of the total value of processed fisheries products for the Gulf, a wide range of other species and product forms are processed in the region. Some species, such as hard clams (*Mercenaria mercenaria*) and oysters (*Crassostrea virginica*), are produced in nearshore aquaculture facilities. Some product forms are also of local significance, such as smoked fish and mullet (*Mugil cephalus*) roe in Florida. In addition to fisheries products obtained from the Gulf of Mexico, finfish and shellfish obtained from other regions of the nation, as well as foreign sources, are processed in the Gulf region. Overall, the Gulf processing and wholesaling industry serves as an important component of product distribution in the U.S. seafood market.

The seafood processing and wholesaling sector also serves as an important source of employment for Gulf region coastal communities. In 1992 approximately 13,000 people were employed annually by these sectors in the Gulf states (fig. 3.8), with the majority of jobs tied to the processing sector. Many coastal communities are dependent on processing and wholesaling because they are economically undiversified with few employment alternatives for their populations (Jacob et al., 2002). However, the employment associated with the industry has been declining in the Gulf region. Employment increased from 12,600 in 1992 to about 16,000 in 1995, was relatively steady through 1997, and then declined dramatically along with the decline in the number of processing plants. The 2002 employment level was almost 37% lower than in 1997.

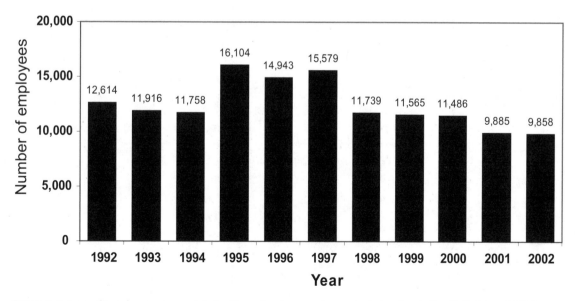

Figure 3.8. Annual employment associated with seafood processing and wholesaling in the Gulf states, 1992 to 2002. Note: All of these figures are for Texas, Louisiana, Mississippi, and Alabama. Source: NOAA, 2003 and 2004.

Economic Activity Associated with Commercial Seafood Production   The commercial fishing and seafood processing/wholesaling sectors in the Gulf region generate considerable economic activity within coastal communities and states, but no recent studies have attempted to measure their collective economic values. Some studies of local impact do exist, however. The commercial seafood industry in Texas reportedly generated about 30,000 jobs and $317 million in payrolls during 1989 (Haby et al., 1993). The industry reportedly generates 29,000 jobs in Louisiana, while contributing $2.6 billion in economic impact and $100 million in taxes to the state's economy (Southwick, 2005). Similarly, 10,600 jobs were created in Mississippi, along with $900 million in economic impact and $42 million in taxes (Posadas, 2005). No statewide estimates exist for Alabama or Florida, although recent studies have shown that the seafood industry in Florida's Lee County, which encompasses a portion of the southwest coast bordering the Gulf, generates $55 million, while Monroe County, which includes the Florida Keys and a lower portion of the southwest coast, generates $160 million in impact to the local economies (Center for Economic and Management Research, 1995; Adams et al., 2002).

Marine Sportfishing

Marine sportfishing is another regional industry dependent on a healthy Gulf of Mexico ecosystem that is an important source of jobs and earnings for many coastal communities. Sportfishing is also a source of recreational activities for many coastal residents and tourists.

Economic Activity Associated with Marine Sportfishing   The economic impact associated with the fishing stems from sales, earnings, jobs, business taxes, and other related activities (American Sportfishing Association, 2005). Associated retail sales have been estimated in excess of $4.3 billion annually (table 3.3). Numbers associated with the Gulf alone are not readily available, but marine sportfishing in Florida as a whole generates more retail sales expenditures and jobs than for all other Gulf region states combined.

Total Gulf fishing expenditures contribute over $2.1 billion in wages and salaries to Gulf region economies and create over 87,000 full-time equivalent jobs, which is estimated to be more economic activity than is associated with any other federal Regional Fishery Management Council area in the United States.

Table 3.3. *Economic Activity from Marine Sportfishing in the Gulf Region, 2001*

State	Retail Sales	Economic Output	Wages & Salaries	Jobs	Sales & Fuel Taxes	State Income Tax	Federal Income Tax
Texas	622	1,328	339	13,322	39	NA	56
Louisiana	410	746	179	7,786	23	4	27
Mississippi	50	98	23	1,003	4	1	2
Alabama	236	464	110	5,477	13	4	11
Florida*	2,987	5,432	1,482	59,418	172	NA	240
Total	**4,305**	**8,068**	**2,133**	**87,006**	**251**	**9**	**336**

Source: NOAA, 2005b.

Units of 1 million (except Jobs, which is expressed in numbers of individual jobs).

* Values for Florida may reflect values associated with both coasts.

Table 3.4. *Marine Sportfishing Anglers and Trips in the Gulf and Other U.S. Regions, 2004*

Region	Non-Resident Anglers	Resident Anglers	Total	Trips
Gulf Region				
Florida	2,542	2,074	4,616	16,616
Alabama	398	407	805	2,048
Louisiana	207	895	1,102	4,810
Mississippi	54	224	278	1,109
Texas	32	426	458	1,040
Total Gulf	3,233*	4,026	7,259*	25,623
South Atlantic	-	2,621	-	20,778
Other	-	2,092	-	28,480
Total United States	-	**8,739**	-	**74,881**

Source: NOAA, 2005b.
Units of one thousand.
* NMFS suggests not adding non-resident anglers across states due to the potential for double counting. The totals presented for non-residents within the Gulf region incorporate that potential source of error.

The charter and party boat industry, which is distinct from the small guide boat industry, is an important component of the overall marine sportfishing industry within the Gulf region, with a significant impact on the economy there. Stoll et al. (2002) found that the charter and party boat operations in the Gulf region generated $149.5 million in economic output, $68.5 million in incomes, and 3,487 jobs within the overall Gulf of Mexico regional economy. These economic activity estimates do not include values deriving from other types of marine boating activities that are not associated with sportfishing.

Numbers of Anglers and Trips In 2004, the Gulf of Mexico region had 7.3 million anglers who took 25.6 million fishing trips, which was more than in any other single region in the United States (table 3.4). These figures are approximately one-third of U.S. totals and about two-thirds of the total for the combined Gulf of Mexico and South Atlantic region. Florida accounts for the largest share of marine sportfishing anglers (64%) and reported trips (65%), while Louisiana reported more saltwater trips than both Mississippi and Alabama combined. The total number of marine sportfishing anglers and trips for the United States, excluding Texas, is estimated via the Marine Recreational Fishery Statistics Survey administered by the National Marine Fisheries Service (NOAA, 2004). Information for Texas anglers is collected through a separate survey done just for that state (Texas Parks and Wildlife Department, 2005).

About half of the participants in marine sportfishing activities in the Gulf region in 2004 were not residents of the states from which they departed (table 3.4). These non-residents are responsible for a large portion of the total marine sportfishing economic activity and are an important source of new revenue for the economies of many coastal communities in the Gulf region. Non-resident expenditures in local economies are spent and respent by local businesses, leading to substantial economic impact for coastal economies.

Of the total number of trips taken by marine sportfishing anglers in the Gulf region during 2003, 61% were taken in private or rental boats, while 31% were taken at shoreside locations, such as beaches, bridges, or piers. The remaining 8% were taken in charter or party boats and vessels (NOAA, 2005b).

Merchant Shipping

Waterborne commerce represents a fourth major industry that is directly associated with and dependent upon the Gulf of Mexico. Major shipping lanes run through the Gulf as well as the protected Intracoastal Waterway system. Waterborne commerce generates economic activity in coastal communities via the associated maritime port facilities, which provide an important source of employment for many small and large coastal communities.

Merchant Shipping Activity The Gulf region has more commercial cargo-handling facilities than any other region of the nation (DOT, 2000), as shown in table 3.5, and thirteen of the country's top ports are located there. Of these, Houston and New Orleans have more port calls, or off-loading events (DOT, 2002), than any other ports. In addition, 70% of all U.S. waterborne commerce, which is measured in ton-miles of shipping, occurs in the Gulf, and 50% of the total U.S. waterborne volume is handled in Gulf ports, though this represents only 19% of the industry's total national value. Sixty percent of all waterborne petroleum and petroleum-based products shipping occurs in the Gulf, which includes a substantial amount of shipping along the Mississippi River corridor into the Gulf. Most of this Mississippi activity—about 93%—occurs between Baton Rouge, Louisiana, and the Gulf.

The volume and value of waterborne commerce has been increasing in the Gulf region in recent years (fig. 3.9), with combined import and export volume rising from 547 million short tons in 1998 to 596 million short tons in 2003 (DOT, 2005). The nominal value of these shipments has increased from $101 billion to $152 billion in the same period. In 2003 the largest share of the total shipping tonnage (73%) and nominal value (66%) was for imports. Recent increases in waterborne commerce transportation in the Gulf have been over twice the national average.

Table 3.5. *U.S. Ports Handling More Than 10 Million Tons in 1997*

Coastal Region	Number of Terminals	Percent of Total	Number of Berths	Percent of Total
Gulf Region	484	25.3	786	24.9
North Atlantic	421	22.0	761	24.1
Great Lakes	340	17.8	483	15.3
North Pacific	249	13.0	365	11.6
South Pacific	223	11.6	414	13.1
South Atlantic	197	10.3	349	11.0
Total	**1,914**	**100.0**	**3,158**	**100.0**

Source: DOT, 2000.
Includes commercial cargo handling facilities with a minimum depth alongside of 25 feet for coastal ports and 18 feet for Great Lakes ports.

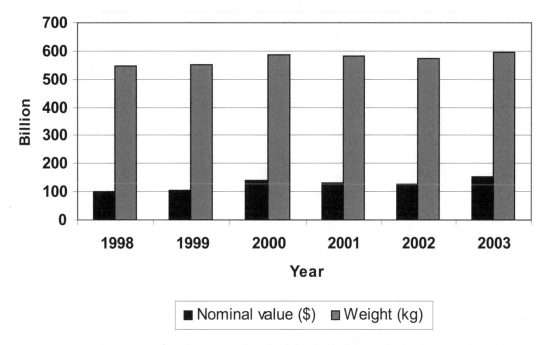

Figure 3.9. Waterborne commerce (imports and exports) for the Gulf Coast district. Source: DOT, 2005.

Economic Activity Associated with Merchant Shipping and Port Facilities Merchant shipping and associated port facilities are important contributors to local Gulf economies. Though no single study provides an aggregate estimate of these industries' economic activities and values, descriptive information for individual ports provides some insight (International Trade Data System, 2006). The Port of New Orleans is the world's largest wharf facility and creates 107,000 jobs and $1.3 billion in earnings annually. The Houston port facilities generate approximately 288,000 jobs and $10.9 billion in revenues. The South Louisiana Port, which stretches over fifty miles upriver from New Orleans, is the third largest port in the world (in terms of tonnage) and handles 15% of the total U.S. export volume. The ports of Freeport and Corpus Christi, Texas, generate 30,000 and 40,000 jobs, respectively, and create $7.1 billion in economic impact and $1.3 billion in revenue, respectively. The Alabama State Docks support 118,000 jobs and create $3.0 billion in economic impact for the Alabama economy. Other ports in the region, such as Tampa and Pensacola, Florida, and Brownsville and Port Arthur, Texas, also contribute to the employment and economic well-being of the local, state, and regional economies.

Cruise Industry

The cruise industry is another major source of economic value emanating from Gulf region port facilities (table 3.6). Each state within the Gulf region receives some economic benefit from the cruise industry.

In 2004 Florida had cruise-related expenditures of $5.2 billion, or 35% of the U.S. total, and total incomes of $4.6 billion, and the industry generated 129,100 jobs. Although these numbers reflect the industry on both Florida coasts, the economic importance of the industry is difficult to overlook. Texas and Louisiana reported expenditures of $709 million (4.5% of the U.S. total) and $208 million, respectively. Texas had about 13,800 cruise-related jobs, while Louisiana had roughly 5,000.

Table 3.6. *Economic Activity Associated with the Cruise Industry in the Gulf Region, 2004*

State	Rank Among All U.S. States	Expenditures	Total Incomes	Total Employment
		$ millions		
Florida (both coasts)	1	$5,157	4,554	129,099
Texas	5	709	578	13,817
Louisiana	14	208	152	5,046
Alabama	21	77	38	985
Mississippi	36	24	12	407

Source: International Council of Cruise Lines, 2005.

Table 3.7. *Cruise Ship Passenger and Trips for the Gulf of Mexico Ports, 2003*

Port	Number of Passengers	Trips
Galveston	754,364	203
Gulfport	27,574	4
Houston	25,638	8
New Orleans	725,439	176
Tampa	834,945	217

Source: DOT, 2005.
Units of individual passengers and trips.

Limited information is available about the economic activities associated with the cruise ship industry (DOT, 2005). Within the Gulf region, Tampa, Galveston, and New Orleans are the most important cruise ship ports (table 3.7). During 2003 approximately 835,000 passengers departed from Tampa on 217 cruise trips. That same year approximately 754,000 and 725,000 passengers departed on cruise ships from Galveston and New Orleans, respectively. Significantly fewer passengers departed from Gulfport, Mississippi, and from Houston.

Maritime Vessel Construction

A number of Gulf region shipyards are actively engaged in producing vessels intended for the commercial maritime industry. Other manufacturers produce vessels and boats intended for the recreational industry, but this sector of the industry is not addressed here. The commercial shipyards within the Gulf region include those identified as major, second-tier, small, aluminum, and inactive/occasional shipbuilders. Of these categories, the most active during 2000–2004 were the shipyards building aluminum ships, followed closely by second-tier and small shipbuilding companies (fig. 3.10). During this period, the total number of vessels constructed per year declined from 150 in 2001 to 116 during 2004 (Colton, 2005). Related declining trends were seen for most types of shipbuilders.

The reported sales for the commercial shipbuilding industry in the Gulf region during 2001 were about $2.2 billion, which represented 56% of total sales for the U.S. industry (LECG, 2002). These sales created economic impacts of $4.0 billion within the U.S. economy, created approximately 54,000 jobs, and generated $3.4 billion in personal incomes. Within the Gulf region Louisiana accounted for the largest share of sales at 45%, economic impacts at 38%, job creation at 38%, and personal incomes at 38% of the Gulf region totals (table 3.8).

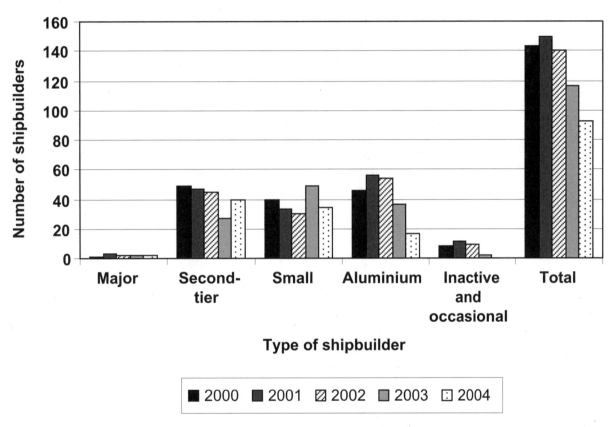

Figure 3.10. Number of ships built in the Gulf states from 2000 to 2004 (in units). Source: Colton, 2005.

Table 3.8. *Economic Activity Associated with Commercial Shipyards in the Gulf, 2001*

State	Sales	Economic Impact	Jobs	Personal Incomes
Alabama	276	513	6,967	439
Florida	202	473	6,359	404
Louisiana	969	1,515	20,756	1,295
Mississippi	386	616	8,430	527
Texas	333	853	11,437	729
Total	**2,166**	**3,970**	**53,950**	**3,394**

Source: LECG, 2002.
Units of one million dollars (except jobs, which are given as total jobs created).

Marine Recreational Activities The broadly defined coastal tourism industry offers a wide range and variety of marine-related recreational activities enjoyed by residents and non-residents within all the states of the Gulf region. Marine recreational activities include beach visitation, swimming, snorkeling, surfing, personal watercraft use, kayaking, scuba diving, and many others. A recent study by U.S. Department of Commerce's National Ocean Service identified nineteen activities, including marine sportfishing (Leeworthy and Wiley, 2001). The study estimated the percentage of the total U.S. population that participated in one or more of these activities, and in what state they did it. Of the U.S. coastal states, Florida had the largest participation rate at 10.7%, indicating that this percentage of the total U.S. population participated in a marine recreational activity in Florida (both coasts). Participation rates for other Gulf states in the study included Texas, with 3.0% participation, Alabama with 1.2%, Louisiana with 1.1%, and Mississippi with 0.9%. The total number of recreational participants within the Gulf states was estimated at 34.7 million.

Unfortunately, no comprehensive studies exist that provide insight into the economic values by state associated with each of the various types of marine recreational activity. However, participation by individuals who do not live within the coastal county corridor generates true economic impact to the coastal counties. Leeworthy and Wiley (2001) found that across all activity types and U.S. states, 37% of participants did not reside in the coastal county or counties where their activity occurred. Though the actual economic impact is not known, that such economic impact occurs is without question.

Conclusion

The Gulf of Mexico serves as a reservoir of economic value upon which many marine-related industries within the region are dependent. Some of these industries, such as oil and gas production and commercial fishing, are engaged in the physical extraction of resources for commercial purposes. Other industries depend on the Gulf for recreational activities such as boating, beach visitation, ecotourism, and diving. Finally, industries such as merchant shipping and ports and marinas are dependent on the logistical value of the Gulf. The economic values associated with these many marine-related industries and activities are significant. No single study assessing the overall economic value associated with the Gulf of Mexico has yet been conducted, but the information in this chapter provides economic snapshots of several key industries and activities that collectively offer some idea of the magnitude of that overall value.

Though its drawing power is not directly addressed in this chapter, the coast of the Gulf of Mexico is also a natural resource that attracts people, increasing the pace of coastal development because people wish to live and work near the water, beaches, and estuaries. Increasing demands by such development and by industries for the Gulf of Mexico's coastal and marine resources reveal a need for more effective resource management. Better awareness of the changing uses of Gulf resources, and the economic values associated with these uses, is essential for enabling resource managers, coastal developers, and individual users to make wise decisions that will ensure the sustainable utilization of this truly unique natural resource.

References

Adams, C., D. Mulkey, and A. Hodges. 2002. Economic Importance of the San Carlos Island Shrimp Processing Industry to the Lee County Economy. In D. Letson and J.

W. Milon (eds.), *Florida Coastal Environmental Resources: A Guide to Economic Valuation and Impact Analysis*. SGR-124. Gainesville: Florida Sea Grant Program.

Adams, C. M., E. Hernandez, and J. Cato. 2005. The Economic Significance of the Gulf of Mexico Related to Population, Income, Employment, Minerals, Fisheries and Shipping. *Ocean & Coastal Management* 47: 565–80.

American Sportfishing Association. 2005. Economic activity data related to marine recreational sportfishing. http://www.asafishing.org.

Applied Technology Research Corporation. 1999. LA Gulf of Mexico Impacts. Prepared for the Louisiana Mid-Continent Oil and Gas Association. Baton Rouge, La..

Baker Hughes, Inc. 2006. North American Rig Count. Rotary rig count, offshore Gulf of Mexico by state. http://www.bakerhughes.com/investor/rig/rig_na.htm.

Center for Economic and Management Research (CEMR). 1995. Economic Impact of Commercial Fisheries in the Florida Keys: Case Study, Florida Keys National Marine Sanctuary Draft Management Plan. Report under contract to the Monroe County Commercial Fishermen, Inc. Tampa, Fla.: Center for Economic and Management Research, University of South Florida.

Colton, T. 2005. Shipbuilding data. The Colton Company. http://www.coltoncompany.com/shipbldg/ussbldrs/postwwii/shipyards.htm.

Darby, K. A. R., D. E. Dismukes, and S. E. Cureington. 2006. Hurricanes and Energy Infrastructure in the Gulf of Mexico: Impacts and Challenges. Center for Energy Studies, Louisiana State University. http://www.searchanddiscovery.net/documents/2006/06086gcags_sec_abs/images/abstract.darby.et.al.pdf.

DOT (U.S. Department of Transportation). 2000. U.S. Economic Growth and the Marine Transportation System. Washington, D.C.: U.S. Department of Transportation, Office of Ports and Domestic Shipping, U.S. Maritime Administration. 22 pp.

———. 2002. Maritime Research and Technology Development: A Report to Congress. Washington, D.C.: U.S. Department of Transportation, U.S. Maritime Administration.

———. 2005. Merchant shipping volumes, value and commercial cruise ship industry data for the Gulf region. Washington, D.C.: U.S. Maritime Administration. http://www.marad.dot.gov/statistics/usfwts/index.html.

Energy Information Agency. 2006. The Impact of Tropical Cyclones on Gulf of Mexico Crude Oil and Natural Gas Production. http://www.eia.doe.gov?emeu/steo/pub/pdf/hurricanes.pdf.

French, L. S., G. E. Richardson, E. G. Kanzanis, T. M. Montgomery, C. M. Bohannon, and M. P. Gravois. 2006. *Deepwater Gulf of Mexico 2006: America's Expanding Frontier*. OCS Report MMS 2006-022. U.S. Department of the Interior, Minerals Management Service, Gulf of Mexico Region. New Orleans. 148 pp.

Haby, M., R. Edwards, A. Reisinger, R. Tillman, and W. Younger. 1993. *The Importance of Seafood-Linked Employment and Payroll in Texas*. TAMU-SG-93-503. Texas Marine Advisory Service, Texas Sea Grant Program. College Station. 10 pp.

Hiett, R. L., and J. W. Milon. 2001. *Economic Impact of Recreational Fishing and Diving Associated with Offshore Oil and Gas Structures in the Gulf of Mexico*. MMS Publication 2002-010. U.S. Department of the Interior, Minerals Management Service, Gulf of Mexico Region. New Orleans, La. 98 pp.

International Council of Cruise Lines. 2005. The Cruise Industry—2004 Economic Summary. Arlington, Va. http://www.iccl.org/.

International Trade Data System. 2006. Unpublished economic activity data for various Gulf maritime ports. Washington, D.C. http://www.itds.treas.gov/ports.html.

Jacob, S., M. Jepson, C. Pomeroy, D. Mulkey, C. Adams, and S. Smith. 2002. Identifying Fishing-Dependent Communities: Development and Confirmation of a Protocol. Silver Spring, Md.: National Oceanic and Atmospheric Administration, National Marine Fisheries Service, Marine Fisheries Initiative Program.

LECG. 2002. The Economic Contribution of the U.S. Commercial Shipbuilding Industry. Washington, D.C.: Prepared for the Shipbuilders Council of America by LECG. 49 pp.

Leeworthy, V. R., and P. C. Wiley. 2001. Current Participation Patterns in Marine Recreation. Special Projects. Silver Spring, Md.: National Ocean and Atmospheric Administration, National Ocean Service. 53 pp.

MMS (Minerals Management Service). 2006. Annual Summary of Production for the

Entire Region, Gulf of Mexico Region, Region Production Years 1953–2006. U.S. Department of the Interior, Minerals Management Service, Gulf of Mexico OCS Region. http://www.gomr.mms.gov/homepg/pubinfo/repcat/product/Region.html.

NOAA (National Oceanic and Atmospheric Administration). 2003. Fisheries Statistics of the United States, 2003. Current Fisheries Statistics no. 2002. Silver Spring, Md.: National Oceanic and Atmospheric Administration, National Marine Fisheries Service.

———. 2004. Fisheries Statistics of the United States, 2003. Current Fisheries Statistics no. 2003. Silver Spring, Md.: National Oceanic and Atmospheric Administration, National Marine Fisheries Service. http://www.st.nfms.gov/st1/.

———. 2005a. Unpublished seafood processing volume and value for the Gulf region. Office of Economics and Statistics. Silver Spring, Md.: National Oceanic and Atmospheric Administration, National Marine Fisheries Service.

———. 2005b. Marine Recreational Fisheries Statistics Survey data. Silver Spring, Md.: National Oceanic and Atmospheric Administration, National Marine Fisheries Service. http://www.st.nmfs.gov/st1/.

Plater, J. R., J. Q. Kelley, W. W. Wade, and R.T . Mott. 1999. *Economic Effects of Coastal Alabama and Destin Dome Offshore Natural Gas Exploration, Development, and Protection.* OCS Study Minerals Management Service 2000-044. U.S. Department of the Interior, Minerals Management Service, Gulf of Mexico OCS Region, New Orleans, La. 219 pp.

Posadas, B. 2005. Economic Impact of the Mississippi Seafood Industry at the Year 2003. Biloxi: Mississippi Sea Grant Extension Program, Mississippi State University. 2 pp.

Scott, L. C. 2002. *The Energy Sector: Still a Giant Economic Engine for the Louisiana Economy.* Prepared for the Louisiana Mid-Continent Oil and Gas Association. Baton Rouge, La.

Southwick, R. 2005. The Economic Benefits of Fisheries, Wildlife and Boating Resources in the State of Louisiana. Alexandria, Va.: Southwick Associates. 38 pp.

Stoll, J., W. Milon, R. Ditton, S. Sutton, and S. Holland. 2002. The Economic Impact of Charter and Party Boat Operations in the Gulf of Mexico. *Proceedings of the Gulf and Caribbean Fisheries Institute* 53: 318–31.

Texas Parks and Wildlife Department. 2005. Unpublished Texas marine sportfishing angler and commercial license data. Austin: Texas Parks and Wildlife Department, Coastal Fisheries Division.

4

The Changing Coastal and Ocean Economies of the United States Gulf of Mexico

JUDITH T. KILDOW, CHARLES S. COLGAN, AND
LINWOOD PENDLETON

Introduction

A comprehensive strategy is needed to protect and nurture the Gulf of Mexico's riches. Public focus—and that of the government and academics—has been largely locked on the devastation from the 2005 hurricanes. But stresses on the rich natural resources of this special area have been intensifying for many years. These stresses have been felt especially in the degradation and shrinkage of wetlands and the decline of fisheries. Declining water quality, both fresh water for drinking and seawater, has exacerbated the situation.

The economies of the Gulf states are inextricably linked to the quality and values of the Gulf's natural resources. Recent reviews of scientific studies, management practices, and the availability of information about economics and natural resources have opened new windows for developing effective strategic plans for protecting these resources. Government and the private sector can work together to use this new information to create a multitiered paradigm with a positive effect on coastal resource management for years to come. Several activities indicate that this shift is under way, including creation of the Gulf of Mexico Alliance (Florida Department of Environmental Protection, 2007) and a new cooperative program in Louisiana known as "A Place Called America's WETLAND" (America's Wetland, 2008). The Gulf of Mexico Alliance is a partnership among states bordering the Gulf, supported by government agencies such as the Environmental Protection Agency (EPA) and the National Oceanic and Atmospheric Administration (NOAA), aimed at sharing science, expertise, and financial resources in order to protect the complex Gulf of Mexico ecosystem better. The Louisiana program is a cooperative venture between the state's Department of Culture, Recreation & Tourism, the America's WETLAND Campaign to Save Coastal Louisiana, the Shell Oil Company, and local leaders and tourism bureaus, designed to promote ecological tourism.

Ecosystem-based management, which is dependent on such alliances and collaborations, is the new mantra. It drives U.S. coastal and ocean policies called for in the report of the U.S. Commission on Ocean Policy (U.S. Commission on Ocean Policy, 2004). Ecosystem-based management is already embedded in current priorities for NOAA, such as the Science for Ecosystem-based Management Initiative, which supports research on the ecological interactions and processes required to sustain healthy ecosystem structures in environments that support fish and fisheries (NOAA Fisheries Service, 2006).

This new shift in emphasis results from the growing realization that the health of the Coastal Economy, or the portion of a region's economy tied geographically to the ocean, is critically important along the Gulf and elsewhere in the world. The combination of harsh storms, widespread effects of El Niño on weather, depletion of fish stocks, marine pollution, beach closures, and intensified human encroachment on once-wild areas provide impetus for a change in strategic thinking.

But changing the traditional paradigm is difficult because our governance system has historically been founded on political jurisdiction–based accountability, not environmental boundaries. Additionally, the health of the economy has not been linked effectively with the health of the environment; rather, economy and environment have too often been placed in opposition.

If the economy and environment of the Gulf of Mexico coast are to grow together in a healthy, sustainable way, it is essential that all who influence the future of the area have the best possible information. Indicators of change are important because they reveal trends in how we use and impact the coastal environment as well as how much we value the natural resources of this area. Such data can be used to create indices of environmental and habitat health, coastal vulnerabilities, and economic vitality, which will allow planners to implement meaningful and enduring strategies for resource management.

A foundation exists on the Gulf Coast from which strategies can be launched to change from the classic to a new management paradigm. That foundation is the strong association of Gulf states with the area's natural systems. Scientists have developed comprehensive time-series data that track changes in the Gulf environment, including the quality of coastal waters, health of fisheries, and the state of wetlands (Coastal America, 2007; NOAA Coastal Services Center, 2008; NOAA Fisheries Service, 2008). But parallel information of similar quality concerning the economy has been lacking.

Now the National Ocean Economy Project (NOEP) has built an extensive database of time-series information that tracks changes in human uses and activities in the coastal environment. This NOAA-sponsored program tracks the growth and decline of ocean-dependent industries, the rapidly changing uses of coastal lands, demographic patterns, housing, the volatility of the productivity of coastal resources, and even the non-market values of recreational and natural assets.[1] In short, NOEP is cataloging and assessing what has been and is happening over time along the entire American coastline and adjacent oceans and is making this information publicly available via the Internet.

The NOEP is a United States effort that has created a template and system to describe and assess changes in the nature and value of the nation's coasts and coastal oceans, including all coastal and Great Lake states. It describes two separate yet overlapping market economies tied to the oceans and designated as the Coastal Economy and the Ocean Economy.

The Coastal Economy encompasses all Super Sectors, as defined by the Bureau of Labor Statistics, based on the North American Industrial Classification System (NAICS). Key geographical units for which information is available are: (1) *coastal counties*—those within the Coastal Zone Management programs of individual states, (2) *watershed counties*—as defined by the U.S. Geological Survey, and (3) *inland areas*—the regions of each state not classified as watershed or coastal. The Coastal Economy includes all economic activities in coastal areas supporting coastal populations, not just coastal-dependent activities. The Ocean Economy, on the other hand, consists of six industrial sectors directly dependent on the ocean as identified by the NAICS. These are Maritime Transportation, Ship & Boat Building and Repair, Living Marine Resources, Offshore Minerals, Coastal Construction, and Coastal Tourism & Recreation.

The first section of this chapter provides a glimpse of the size and scope of the Coastal Economy for the American side of the Gulf Coast. Comparisons of sectors, years, and states using estimates of changes in wages, employment, and gross state product (GSP) over the past fourteen years are provided as indicators of strengths and weaknesses of the economies along the coast.[2] Data on population and housing growth as well as some natural resources information complete the picture of the Coastal Economy from the market perspective. The second part of the chapter addresses the six sectors of the Ocean Economy, comparing geographies and sectors to indicate how the Gulf Ocean Economy ranks. Comparisons are made among the sectors, locally, among other coastal states and regions, and nationally. This part of the chapter also provides forecasts for the future of these sectors in the Gulf of Mexico using information from the past fourteen years. The third section of the chapter provides insights into the non-market values of resources and assets of the Gulf Coast, demonstrating the limits and powers of estimating values not measurable in the marketplace and providing some estimates of rarely considered values.

Coastal Economy of the Gulf of Mexico

The American coastal counties ringing the Gulf of Mexico measure approximately 49,131 square miles, which is 1.3% of total U.S. land area and a little more than 7% of the total area of all American coastal counties (fig. 4.1). Although constituting only a modest portion of land area, these counties make a substantial contribution to the gross domestic product (GDP).

Scale and Scope of the Coastal Economy In 2004, with only 1.3% of the land, Gulf coastal counties contributed 4.7%, or $496 billion, of the nation's $10.7 trillion GDP. All U.S. coastal counties contributed a total of $5 trillion, or 47.5% of the total U.S. GSP (table 4.1).

Furthermore, to show the larger importance of the coast to the U.S. economy, the contribution of all coastal states was 83% of U.S. GDP, or $8.9 trillion in 2004.[3] The Gulf states contributed $1.7 trillion, or about 16% of this amount, or 19% of total coastal state contributions. Coastal areas contribute a disproportionately large share of the U.S. economy, though they encompass only a very small percentage of U.S. land (table 4.1).

In addition, the Gulf coastal counties produced 27% of the employment, 26% of wages, and 26% of the GSP of the total economy of the Gulf states. Florida and Texas combined represent 69% of the land area of the Gulf coast counties, Louisiana follows with 22%, Mississippi with 7%, and Alabama with 6% of the coastal county land, as measured by counties bordering the Gulf (fig. 4.1).

Population and Housing along the Gulf Coast With concern that a disproportionate segment of the U.S. population is moving to vulnerable coastal areas, it should be noted that between 1990 and 2004, the two states with the highest population and housing numbers experienced higher growth rates inland than along the coast. Texas saw a 35% inland population growth rate, compared to 26% on the coast, while inland Florida's population grew by 42% compared to 32% on the coast (table 4.2). In contrast, the other states continued to experience higher population and housing growth along the coast than inland from 1990 to 2004, except for Louisiana, where population growth was slightly higher on the coast, but housing grew faster inland. Of the other states, coastal Mississippi experienced the highest growth rate, with a 20% increase in population and a 27% increase in housing (table 4.3).

However, rates of change can be deceptive if the base is very low. It is when the base number is high to begin with, as with Florida and Texas, that the actual numbers tell the real growth story. Texas and Florida, states with the largest

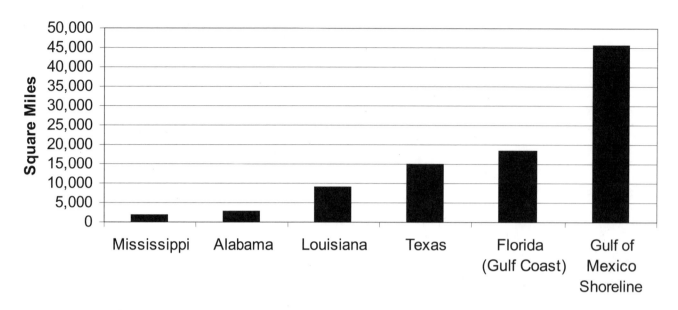

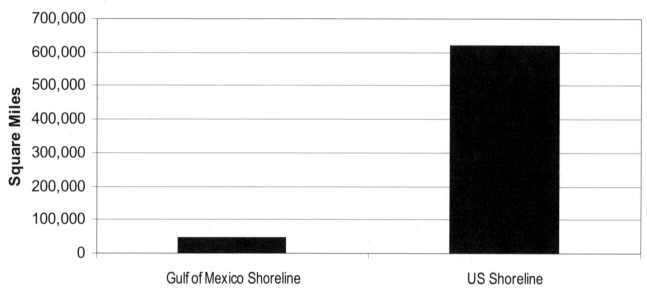

Figure 4.1. Land distribution of coastal counties in the Gulf of Mexico states.

Table 4.1. *Regional Comparison of Gross State Product and Land Area, 2004*

Region	GSP ($ million)	Area (sq. miles)
Gulf Coastal Counties	$496,222.3	49,131.3
Gulf States	$1,697,796.0	456,936.7
U.S. Coastal Counties	$5,063,407.6	678,430.6
U.S. Coastal States	$8,878,097.0	2,002,441.8
U.S. National	$10,662,196.0	3,794,083.1

Source: U.S. Bureau of Labor Statistics.

Table 4.2. *Gulf of Mexico Coastal Population by State, 1990 to 2004*

Region	1990	2004	Growth Rate
Alabama Coastal	476,923	557,227	16.8%
Alabama Inland	3,563,466	3,972,955	11.5%
Florida Coastal*	4,069,679	5,378,241	32.2%
Florida Inland	2,871,728	4,064,281	41.5%
Louisiana Coastal	1,810,034	1,941,296	7.3%
Louisiana Inland	2,411,792	2,574,474	6.7%
Mississippi Coastal	312,368	373,762	19.7%
Mississippi Inland	2,263,107	2,529,204	11.8%
Texas Coastal	4,394,982	5,548,520	26.2%
Texas Inland	12,591,443	16,941,502	34.5%
Gulf Coastal	11,063,986	13,799,046	24.7%
Gulf Inland	23,701,536	30,082,416	26.9%

*Gulf counties only
Source: U.S. Census Bureau.

Table 4.3. *Gulf of Mexico Coastal States Housing Rate of Growth, 1990 to 2004*

Region	1990	2004	Growth Rate
Alabama Coastal	202,153	258,118	27.7%
Alabama Inland	1,468,226	1,800,833	22.7%
Florida Coastal*	2,083,519	2,725,692	30.8%
Florida Inland	1,254,512	1,777,631	41.7%
Louisiana Coastal	740,406	812,964	9.8%
Louisiana Inland	975,835	1,106,894	13.4%
Mississippi Coastal	129,916	165,100	27.1%
Mississippi Inland	880,507	1,056,140	19.9%
Texas Coastal	1,810,379	2,195,246	21.3%
Texas Inland	5,198,620	6,651,482	27.9%
Gulf Coastal	4,966,373	6,157,120	24.0%
Gulf Inland	9,777,700	12,392,980	26.7%

*Gulf counties only
Source: U.S. Census Bureau.

base population and housing, experienced by far the largest coastal and inland population growth during that period. These states were also largely responsible for a higher inland than coastal growth rate for coastal states taken as a whole. Though with differing rates, population and housing rates are increasing in all Gulf states and, as of 2004, keeping pace with each other.

How do Gulf state growth rates compare with other parts of the nation? Patterns appear to differ according to the region. The West Coast has a similar pattern to the Gulf states, with slower growth along the coast in more populated areas. This is possibly due to saturation, with increasingly limited land availability, and land costs escalating beyond the reach of most of the population (National Ocean Economics Program, 2007). Coastal regulations also limit uses in some sensitive areas, such as those with protected estuaries and those with particularly high erosion. In some coastal areas, states have bought large undeveloped areas for preservation and conservation when they were designated for special attributes.

State-Level Coastal Economy For measuring the importance of the Coastal Economy, Louisiana provides a good example of how important a state's coast can be. There, the coastal parishes accounted for 48.8% of total state employment and 50.8% of GSP in 2004. Yet coastal parishes in Louisiana make up only 21% of the total square miles of Louisiana. Clearly, coastal counties are a key part of the state's economy. Combine that intensity of economic activity in a small area with the low-lying geography along the coast, and Louisiana's vulnerability to natural coastal disasters becomes clearly apparent (fig. 4.2). The state's resiliency could be compromised by its exposure to natural hazards of potentially increasing intensity and frequency.

During 2004 the Louisiana parishes with the largest Coastal Economies were Orleans and Jefferson. Orleans Parish contributed 13.3% of Louisiana's employment and 15.4% of the GSP. Jefferson Parish contributed 11.5% of employment and 11.7% of GSP. Again, these areas supporting a disproportionately large part of Louisiana's economy are low-lying, leaving them extremely vulnerable to natural hazards.

The states with the largest coastal county employment growth rates from 1990 to 2004 were Mississippi at 31% and the Gulf Coast of Florida with 38%.

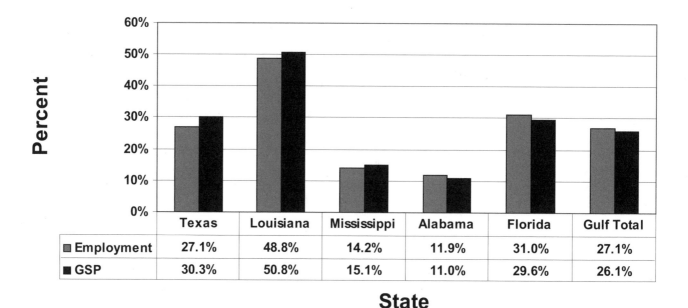

	Texas	Louisiana	Mississippi	Alabama	Florida	Gulf Total
■ Employment	27.1%	48.8%	14.2%	11.9%	31.0%	27.1%
■ GSP	30.3%	50.8%	15.1%	11.0%	29.6%	26.1%

State

Figure 4.2. Gulf coastal county contributions to state employment and gross state product (GSP), 2004. Source: U.S. Bureau of Labor Statistics.

Table 4.4. *Employment Growth in Gulf Coastal Counties, 1990 to 2004*

Region	Employment (1990)	Employment (2004)	Growth Rate
Texas Coastal	2,017,657	2,505,095	24%
Louisiana Coastal	702,625	810,602	15%
Mississippi Coastal	114,189	150,096	31%
Alabama Coastal	172,139	215,200	25%
Florida Coastal*	1,601,303	2,209,215	38%
Gulf Coastal	4,607,913	5,890,208	28%

*Gulf counties
Source: U.S. Bureau of Labor Statistics.

Table 4.5. *Summary of Gulf States Coastal County Economy, 2001 and 2004*

Year	State	Employment	Wages (in billions)	GSP (in billions)
2001	Alabama	215,971	$5.9	$13.1
2004	Alabama	215,200	$6.0	$14.1
2001	Florida*	2,126,933	$62.0	$142.7
2004	Florida*	2,209,215	$67.8	$163.7
2001	Louisiana	812,433	$24.0	$58.7
2004	Louisiana	810,602	$24.7	$62.8
2001	Mississippi	148,169	$4.1	$10.0
2004	Mississippi	150,096	$4.1	$10.4
2001	Texas	2,532,848	$98.7	$227.5
2004	Texas	2,505,095	$97.6	$245.2
2001	Gulf Region	5,835,354	$194.7	$452.0
2004	Gulf Region	5,890,208	$200.2	$496.2
Change		0.9%	2.8%	9.8%

*Gulf counties only
Source: U.S. Bureau of Labor Statistics.

Florida's Gulf Coast employment grew at a far greater rate than the rest of the Gulf states (table 4.4).

While growth appeared robust during the 1990s, the years 2001–2004 saw a slowing of the economies in the Gulf coastal counties. The annual growth rate in employment was 0.3% during the latter years compared to a 2.3% annual growth rate between 1990 and 2000, annual growth for wages of 0.9% compared to a 3.4% growth during the 1990s, and a 3.2% annual growth in GSP (table 4.5) compared to 3.6% during the 1990s. The higher rate for GSP between 2001 and 2004 was possibly due to the rate of change in the prices of oil and gas.

Comparisons of Housing, Population, and Employment It is informative to consider how housing, population, and employment track each other. In California, housing along the coast has not kept pace with either employment or population growth, with employment far outpacing both housing and population, creating a problem with affordable housing in many places along California's coast (National Ocean Economics Program, 2007). The Gulf Coast has a similar situation (fig. 4.3), with employment outpacing both housing and population, but not to the extreme levels seen in California.

Coastal Vulnerability   If one assumes that Gulf coastal areas are the most hazardous for natural disasters, and that assets in those areas are most at risk, a look at the percentage of the Gulf states' population, housing, employment, and land area concentrated on the coasts (fig. 4.4) provides some interesting observations. Florida and Louisiana both stood out in 2004 with the largest percentages of state economy and population residing in coastal counties. However, Louisiana had a much larger ratio of land use to land than Florida. Florida, with its longer coastline, has an almost even ratio with its percentages of overall housing, employment, and population on the state's coast about the same as the percentage of land on the coast. In other words, the state distributes its population and economy more evenly and with less density.

Louisiana and the rest of the Gulf states have a much greater proportion of their assets located on coastal lands, with ratios ranging from 2:1 to 5:1 for population and economy to land. Clearly these states have a propensity for squeezing population and the accompanying economy onto coastal land, no matter how small the available acreage. Looking only at percentage of the state's economy in terms of population, housing, and employment located along the coast, Alabama's coastal counties posed the lowest threat to the entire state economy in 2004, with Mississippi close behind. This would indicate that their less intensive coastal uses made them the least vulnerable of all Gulf states to disaster.

Louisiana's state economy was the most vulnerable in 2004, according to both land use and concentration of population and economy, with 44% of the state employment coming from coastal parishes and 43% of the population,

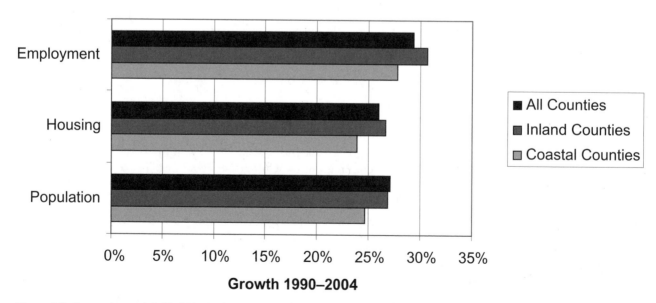

Figure 4.3. Comparisons of Gulf of Mexico housing, population, and employment growth, 1990 to 2004. Sources: U.S. Bureau of Census; U.S. Bureau of Labor Statistics.

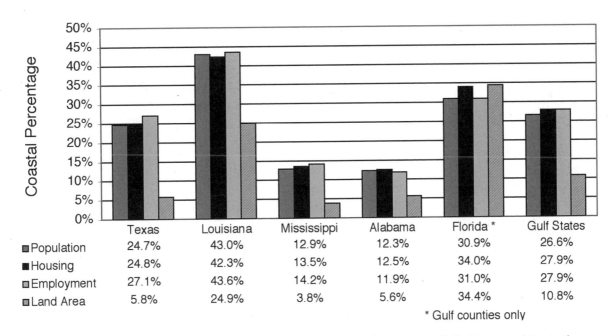

	Texas	Louisiana	Mississippi	Alabama	Florida *	Gulf States
▣ Population	24.7%	43.0%	12.9%	12.3%	30.9%	26.6%
■ Housing	24.8%	42.3%	13.5%	12.5%	34.0%	27.9%
▣ Employment	27.1%	43.6%	14.2%	11.9%	31.0%	27.9%
▣ Land Area	5.8%	24.9%	3.8%	5.6%	34.4%	10.8%

* Gulf counties only

Figure 4.4. Comparison of the size of the coastal county economy for each state in the Gulf of Mexico relative to the percent of land represented by each state's coastal county, 2004. Source: U.S. Bureau of Labor Statistics.

while occupying 25% of the state's land (fig. 4.4). However, another important perspective to consider is the actual value of assets exposed. In 2004 the Texas Coastal Economy was the largest exposed economy (at $245.2 billion in GSP) of all the Gulf states and therefore had the greatest potential for total loss in actual values (table 4.5). Mississippi represented the smallest potential loss value at $10.4 billion in GSP (table 4.5).

Using numbers available immediately after Hurricane Katrina occurred in 2005, the NOEP and the NOAA Coastal Services Center created a map indicating probable losses. The economic data used were for 2003, the most recent year for which data were available prior to 2005. At the time, Gulf coastal states represented approximately 5% of the 2003 U.S. Coastal Economy. While seemingly a small percentage, this is a significant amount of the economy for these states. Nearly half the jobs and more than half the economic output of Louisiana, Mississippi, and Alabama were affected by Katrina. The eighty counties initially identified as most severely damaged (fig. 4.5) had more than two million employees in 2003 and produced an economy of more than $180 billion (National Ocean Economics Program, 2007).

Natural Resources Values Offshore natural resources are important to the economy of the Gulf states but face a number of significant threats. The fishing industry, for instance, is exposed to multiple threats including pollution, wetland destruction and erosion, warming waters, and overfishing. As a result, the industry has experienced some volatility, with declines in some species and stability in others. Using data from the National Marine Fisheries Service (NMFS) for all species caught since 1963 based on recorded landings, landed values, and price per pound, it is possible to identify which species have declined, causing income losses to local areas over time, and which species have contributed to the local, state, and regional economies. One way to identify changes in the fishing industry is to look at the history of the landings and value over time for

Almost half the jobs and more than half the economic output of Louisiana, Mississippi, and Alabama come from the area hardest hit by Hurricane Katrina. In the 80 counties initially identified as most severely impacted by this storm, more than two million employees in 2003 created $180 billion of economic output. Economic activity in this 80-county area represents a fraction of the national economy, but the state and regional impacts are profound.

These data are from the Department of Commerce-sponsored National Ocean Economics Program (NOEP). For more information on the NOEP, see www.oceaneconomics.org.

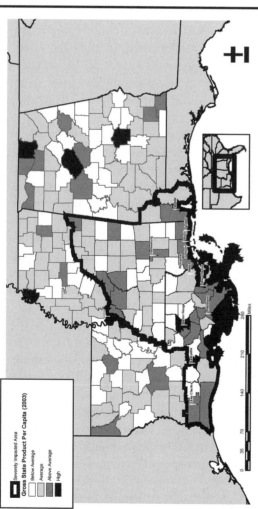

Louisiana

In Louisiana, productivity within the most severely impacted counties is about 80 percent higher than in the rest of the state. The six counties with the highest output per employee were among those counties most severely impacted by the storm.

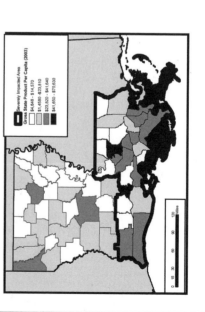

Mississippi

The most severely impacted counties in Mississippi account for 58 percent of the area of that state, 65 percent of the state's employment, and 68 percent of the output. The large geographic extent of the damage accounts for the enormity of impact to this state.

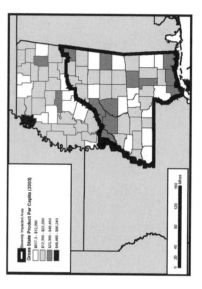

Alabama

The initial disaster declarations identified only three counties in Alabama as severely impacted by Katrina. These three counties, comprising eight percent of the area of the state, includes 12 percent of the employment and 11 percent of the output. Seven other counties were subsequently added to the list of severely impacted counties in this state.

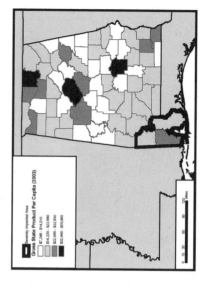

Figure 4.5. Hurricane vulnerability map based on 2003 data for Gulf of Mexico states. Source: NOAA Coastal Services Center.

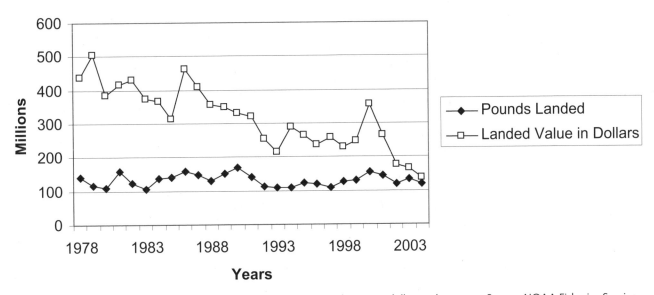

Figure 4.6. Gulf of Mexico brown shrimp harvest and value adjusted to 2000 dollars as base year. Source: NOAA Fisheries Service.

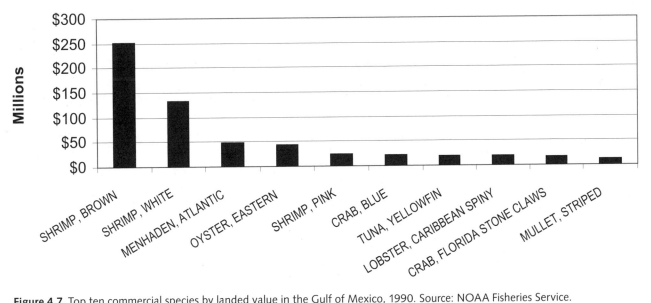

Figure 4.7. Top ten commercial species by landed value in the Gulf of Mexico, 1990. Source: NOAA Fisheries Service.

a single species. For the Gulf region, the history of one of the most popular species, brown shrimp (*Farfantepenaeus aztecus*), graphically depicts the saga of changes in value and catches for a fishery over decades (fig. 4.6; National Ocean Economics Program, 2007). During this time there were also changes in species catch, either due to regulations and restrictions imposed for management or due to declines that led to replacements. There were almost uniform declines in all species caught in both 1990 and 2003, and there was also addition of new species and subtraction of other species during this time (figs. 4.7 and 4.8).

Comparing 1990 with 2003, the top ten species for each year in landed value indicate the changes in productivity and the sustainability of the fisheries as well as changes in markets. The rankings for the top five species have remained relatively stable, but not the bottom ones. For example, striped mullet (*Mugil ceph-*

alus) was removed from the list in 2003, yellowfin tuna (*Thunnus albacares*) dropped in ranking, red grouper (*Epinephelus morio*) was added to the list, and Florida stone crab (*Menippe mercenaria*) increased its rank from ninth to seventh (National Ocean Economics Program, 2007).

Between 1990 and 2003, sharks and yellowfin tuna dropped from the top ten species in landed weight, while Florida stone crab and red grouper were added from 1990 to 2003 (figs. 4.9 and 4.10). With closer scrutiny and over longer periods of time, the depletion of popular species and addition of new ones become even more apparent.

The Gulf region has also been a crucial place for offshore production of oil and natural gas for the United States, and an essential part of the Gulf economy,

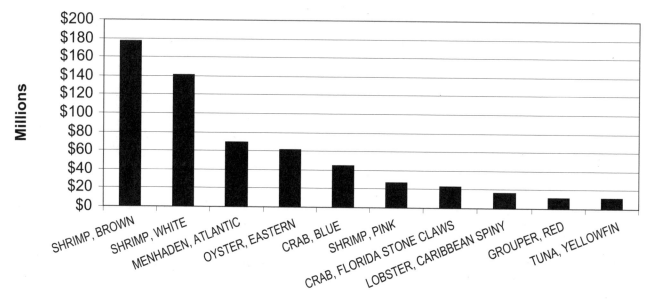

Figure 4.8. Top ten commercial species by landed value in the Gulf of Mexico, 2003. Source: NOAA Fisheries Service.

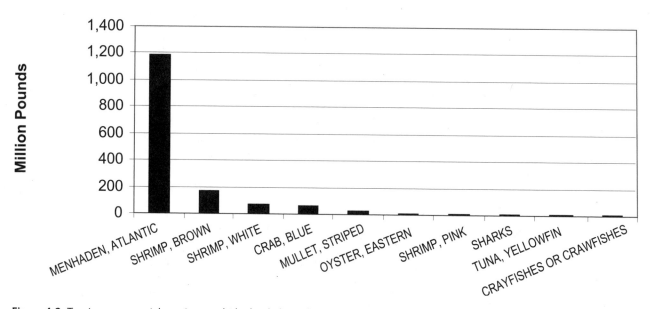

Figure 4.9. Top ten commercial species caught by landed weight in the Gulf of Mexico, 1990. Source: NOAA Fisheries Service.

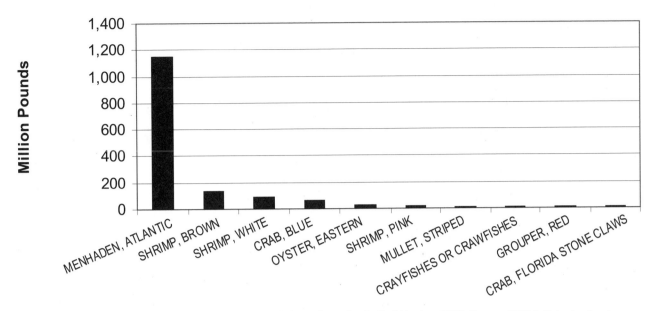

Figure 4.10. Top ten commercial species caught by landed weight in the Gulf of Mexico, 2003. Source: NOAA Fisheries Service.

at least for the states in the western portion of the Gulf. The unprecedented rise in the prices of oil and gas in the past two years is not reflected in the summaries presented here. However, despite the two-year gap in price, the wellhead price for gas, or the value paid at the well excluding cleaning, transportation, and distribution charges that will be required, more than doubled between 1994 and 2004 (fig. 4.11). The first purchase price for oil, which reflects the costs beyond the well, almost quadrupled during the same decade. Of note is the steady decrease in production of natural gas in the Gulf, while oil production increased steadily over that decade (figs. 4.11 and 4.12).

As an indicator of the importance of this natural resource to the nation as well as the Gulf of Mexico states, consider that the total offshore production of oil in the Gulf represents 80% of the total U.S. offshore production, and 22% of total U.S. oil production. The value reflects 81% of the value from offshore oil for the nation. The total offshore production of natural gas in the Gulf represents 97% of the total U.S. offshore production and 19% of the country's total production. Gulf gas was 99% of the U.S. value for offshore and 19% for total U.S. gas production. This is an extraordinary contribution to both the Gulf economy and the U.S. economy (National Ocean Economics Program, 2007). In addition, the revenue generated through wages and jobs from this industry and its support structure in the Gulf is not trivial.

Summary   The resources and the economy of the Gulf of Mexico have changed significantly from 1990 to 2004, as illustrated in a number of ways in this chapter. The information presented here is only a sampling of the diverse Coastal Economy among the Gulf states and of the many measures that can be used to describe changes and values within that economy, such as rates and patterns of productivity; employment, wage, population, and housing growth; and natural resource trends based on geographical measures. The data on the Gulf of Mexico states reveal a growing population and economy that, before the 2005 hurricanes, appeared to be balanced, with the exception that employment, at

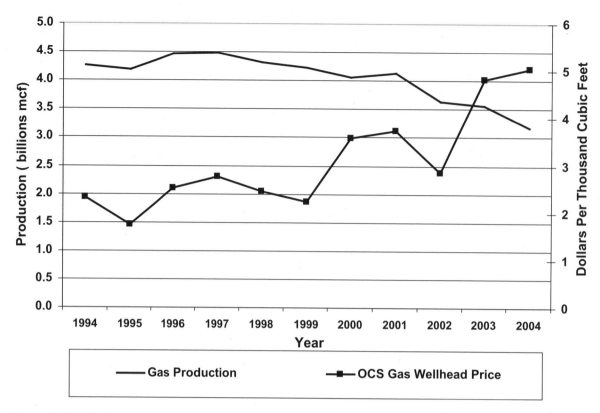

Figure 4.11. Gulf of Mexico Outer Continental Shelf (OCS) gas production and pricing, 1994 to 2004. Source: U.S. Department of the Interior, Minerals Management Service.

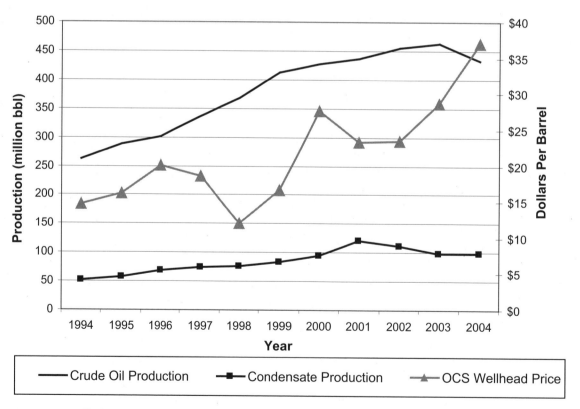

Figure 4.12. Gulf of Mexico Outer Continental Shelf (OCS) oil production and pricing, 1994 to 2004. Source: U.S. Department of the Interior, Minerals Management Service.

28% growth, was slightly outpacing the 25% population growth and the 24% housing growth. Also, the population of inland areas was growing at a slightly higher rate on average, 27%, than in coastal areas, 25%. With a heavy dependence on natural resources, the information provided indicates that fisheries are a volatile industry warranting constant vigilance by managers. The offshore oil and gas industry, with its location and dominance for the nation, remains vulnerable to storms and presents a risk to the U.S. economy. After the 2005 hurricane season, it is apparent that some of the previous rates of growth along the coast in population, housing, and the economy should be reexamined for vulnerability so that the Gulf Coast can gain resiliency and remain robust and productive for generations to come.

Ocean Economy of the Gulf of Mexico in National Perspective

The Gulf of Mexico is both the industrial heartland of the U.S. Ocean Economy and the fastest growing of the major coastal regions in terms of tourism and recreation. These two defining sectors of the Gulf of Mexico's Ocean Economy shape a complex and evolving relationship between the U.S. coastal regions and the Gulf of Mexico that is a key part of the overall relationship between people and the Gulf.

Gulf of Mexico's Ocean Economy This section of the chapter describes the current dimensions and recent history of the Gulf of Mexico Ocean Economy, placing them within the context of the overall U.S. Ocean Economy. The data from this analysis are the most current available and come from an analysis of the Ocean and Coastal Economies of the United States conducted by the NOEP.

For the purposes of this analysis, the Ocean Economy is defined to consist of six major sectors and twenty-three industries (table 4.6). These sectors and industries are defined using a combination of industrial classifications from the North American Industrial Classification System, for data after 2001, and the Standard Industrial Classification, for data prior to 2001. For some industries, such as those in Tourism & Recreation, Construction, and Minerals, an additional defining characteristic of location is used. An establishment is part of the Ocean Economy if it is located in a zip code adjacent to the shore of an ocean or Great Lake (including major bays like Tampa Bay). Locations were determined using confidential establishment-level data from the Bureau of Labor Statistics Quarterly Census of Employment and Wages (QCEW).[4]

Employment data in the NOEP database are for annual average wage and salary, with self-employment excluded. In addition to employment and wages, the NOEP database includes a measure of output for each industry defined as a contributor to GSP. GSP is the state-level analog to the gross domestic product, which is the national measure of the value of goods and services. Each ocean industry establishment's contribution to GSP is set equal to the establishment's wages as a proportion of the total wages paid in each industry (table 4.7).

In 2003, using the NOEP definitions, the Gulf of Mexico Ocean Economy employed nearly 562,000 people and paid wages of more than $13.2 billion, while contributing more than $32 billion to the region's GSP. The largest sector in all three measures is Tourism & Recreation, which is true in other coastal states and the United States as a whole. Tourism & Recreation makes up 71% of Gulf Ocean Economy employment, but only 41% of the contribution to GSP (fig. 4.13). In contrast, the Minerals sector, which is overwhelmingly domi-

Table 4.6. *The Ocean Economy Sectors and Industries*

Sector	Industry	Sector	Industry	Sector	Industry
Construction	Marine Construction*	Tourism & Recreation*	Amusement & Recreation Services	Transportation	Deep Sea Freight Transportation
Living Resources	Fish Hatcheries & Aquaculture*		Boat Dealers		Marine Passenger Transportation
	Fishing		Eating & Drinking Places		Marine Transportation Services
	Seafood Markets		Hotels & Lodging Places		
	Seafood Processing				Search and Navigation Equipment
Minerals	Oil and Gas Exploration and Production*		Marinas		Warehousing*
	Limestone, Sand & Gravel		Recreational Vehicles Parks & Campsites		
Ship & Boat Building	Boat Building & Repair		Scenic Water Tours		
	Ship Building & Repair		Sporting Goods		
			Zoos, Aquaria		

* Included in Ocean Economy if an establishment is located in a shore-adjacent zip code.
Source: Industry groupings determined by the authors and the National Ocean Economics Program based on the North American Industrial Classification System.

nated by oil and gas exploration and the production industry, makes up only 3% of employment but more than 26% of GSP. The Transportation sector is similar to the Minerals sector in that its 13% share of employment is exceeded substantially by a 20% share of GSP. In table 4.7, the category Eating & Drinking Places, not surprisingly, is the largest industry in terms of employment, while Oil & Gas Exploration and Production (Minerals) is the largest industry in terms of GSP.[5]

The composition of the Gulf of Mexico Ocean Economy differs in important ways from that of the U.S. Ocean Economy. For instance, the Gulf of Mexico's Ocean Economy has a smaller share of employment in Tourism & Recreation and larger shares for Construction and Ship & Boat Building (fig. 4.13).

The Gulf has a much larger proportion of its Ocean Economy GSP in Minerals, as would be expected. Compared to the national numbers, the Gulf also has a smaller share of its Ocean Economy employment in Transportation, though that makes up a larger share of its GSP.

The differences between the Gulf of Mexico and the United States (fig. 4.13) point to one of the most important facts about the Gulf Ocean Economy: high proportions of a number of key sectors in the U.S. Ocean Economy are concentrated in the Gulf of Mexico. The proportion of the U.S. Ocean Economy located in the Gulf of Mexico in both 1990 and 2003 is shown in table 4.8. In 2003 the Gulf of Mexico accounted for more than a quarter of the U.S. Ocean Economy employment, slightly less of wages, and slightly more of GSP. Gulf of Mexico employment in the Construction, Living Resources, Minerals, and Ship & Boat Building sectors is a larger proportion of U.S. totals, with marine

Table 4.7. *The Gulf of Mexico and U.S. Ocean Economies, 1990 and 2003*

Sector	Industry	1990			2003		
		Employment	Wages ($M)	GSP ($M)	Employment	Wages ($M)	GSP ($M)
Construction	Total	15,474	$401.70	$730.40	16,205	$671.40	$1,199.30
Living Resources	Fish Hatcheries & Aquaculture	1,288	$19.50	$109.20	1,612	$34.50	$290.60
	Fishing	1,858	$29.40	$135.50	295	$5.70	$22.00
	Seafood Markets	7,111	$84.70	$178.30	2,672	$54.70	$122.80
	Seafood Processing	15,041	$199.80	$503.10	13,055	$271.10	$592.10
	Total	27,163	$355.30	$982.20	18,887	$394.20	$1,138.60
Minerals*	Total	17,817	$621.50	$4,836.80	17,414	$1,109.40	$8,713.20
Ship & Boat Building*	Total	36,750	$906.20	$1,324.90	40,493	$1,547.20	$1,958.20
Tourism & Recreation	Amusement & Recreation Services	7,411	$86.20	$160.80	7,724	$133.10	$264.60
	Boat Dealers	3,292	$68.20	$141.20	4,913	$162.70	$355.10
	Eating & Drinking Places	173,341	$1,546.70	$3,238.70	283,519	$3,771.60	$7,100.80
	Hotels & Lodging Places	72,463	$879.50	$2,039.70	89,398	$1,843.50	$5,060.20
	Marinas	2,503	$42.00	$79.70	3,370	$81.70	$175.00
	Recreational Vehicles Parks & Campsites	893	$11.10	$25.70	1,236	$23.00	$58.20
	Scenic Water Tours	0	$0.00	$0.00	2,530	$56.50	$100.90
	Sporting Goods	487	$7.30	$16.30	983	$27.40	$63.40
	Zoos, Aquaria	435	$5.70	$9.20	2,127	$44.40	$90.20
	Total	261,179	$2,650.80	$5,718.00	398,642	$6,183.90	$13,349.40
Transportation	Deep Sea Freight Transportation	14,769	$545.70	$1,002.90	6,308	$391.90	$1,012.60
	Marine Passenger Transportation	5,519	$128.10	$248.10	8,059	$365.70	$1,014.20
	Marine Transportation Services	35,184	$810.80	$1,478.50	36,209	$1,422.50	$2,500.30
	Search and Navigation Equipment	14,427	$437.10	$579.20	14,199	$912.90	$1,429.90
	Warehousing	1,836	$31.00	$61.80	4,522	$171.70	$347.20
	Total	76,702	$2,064.10	$3,536.40	70,746	$3,324.00	$6,396.80
Ocean Economy	Total	435,086	$6,999.70	$17,128.80	562,387	$13,230.00	$32,755.50

* Individual industries not shown to avoid suppressions.
Source: Based on NOEP industry groupings with data from U.S. Bureau of Labor Statistics.

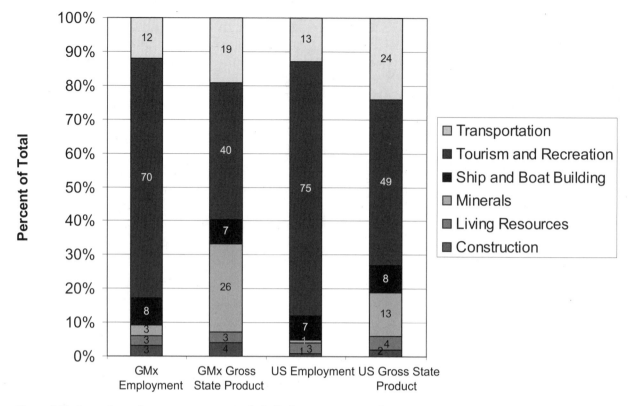

Figure 4.13. Percentage of ocean economy in each Gulf of Mexico (GMx) industry sector, 2003. Source: U.S. Bureau of Labor Statistics.

construction in the Gulf accounting for almost half of national employment in this sector. The proportion of U.S. wages and GSP in the Gulf in 2003 was also substantial in both Construction and Minerals.

From 1990 to 2003, the Gulf of Mexico increased its share of the U.S. Ocean Economy overall from 21.6% of employment to 25.8% of employment, and from 20.9% of GSP to more than 27.4% of national Ocean Economy GSP. From 1990 to 2003, the Gulf's employment shares for Construction, Living Resources, Ship & Boat Building, and Transportation all increased. In Minerals, the Gulf's employment share was largely unchanged, while the share of Tourism & Recreation employment grew slightly. The Gulf's share of national Ocean Economy GSP during this period increased in all sectors except Transportation, where there was a slight decrease. The high proportions of employment and GSP in the national Ocean Economy sectors other than Tourism & Recreation lead to the conclusion that the Gulf is the industrial heartland of the U.S. Ocean Economy, as already mentioned.

Among the individual industries, the Gulf of Mexico stands out for its share of Marine Passenger Transportation employment, which almost doubled between 1990 and 2003, as well as Ship Building & Repair, Seafood Processing, Fish Hatcheries & Aquaculture, and Marine Transportation Services. The Gulf accounts for more than 40% of GSP in Marine Construction, Fish Hatcheries & Aquaculture, and Marine Transportation Services.

Note that while the Gulf of Mexico Ocean Economy's share of the Tourism & Recreation sector was smaller than that for other industries between 1990 and 2003, its importance to the region and its share of the national total are increasing (table 4.8). This point becomes even clearer in table 4.9, where the growth

Table 4.8. *Proportion of U.S. Ocean Economy in Gulf of Mexico, 1990 and 2003*

Sector	Industry	1990			2003		
		Employment	Wages	GSP	Employment	Wages	GSP
Construction	Total	**45.5%**	**37.4%**	**34.0%**	**54.6%**	**47.1%**	**45.6%**
Living Resources	Fish Hatcheries & Aquaculture	39.2%	34.0%	30.1%	34.2%	27.9%	47.0%
	Fishing	15.0%	7.0%	5.9%	5.5%	2.3%	2.3%
	Seafood Markets	19.6%	17.1%	14.5%	20.9%	20.2%	20.7%
	Seafood Processing	26.5%	18.6%	21.0%	32.8%	23.4%	21.7%
	Total	**25.0%**	**17.4%**	**15.7%**	**30.1%**	**21.9%**	**23.3%**
Minerals	Total	**62.7%**	**55.0%**	**54.7%**	**61.0%**	**55.5%**	**57.0%**
Ship & Boat Building	Total	**16.0%**	**13.8%**	**17.2%**	**26.1%**	**22.1%**	**20.5%**
Tourism & Recreation	Amusement & Recreation Services	10.9%	8.0%	7.7%	16.7%	14.5%	17.1%
	Boat Dealers	26.7%	26.3%	25.6%	33.6%	32.4%	32.4%
	Eating & Drinking Places	21.2%	19.5%	17.8%	24.3%	22.3%	22.4%
	Hotels & Lodging Places	23.2%	18.4%	18.0%	26.1%	22.2%	23.7%
	Marinas	23.4%	19.9%	19.5%	22.8%	19.8%	22.5%
	Recreational Vehicles Parks & Campsites	25.5%	23.7%	24.4%	23.8%	22.9%	22.8%
	Scenic Water Tours	NA	NA	NA	27.6%	27.1%	28.1%
	Sporting Goods	8.9%	5.9%	5.3%	14.1%	9.1%	9.7%
	Zoos, Aquaria	10.4%	7.2%	5.3%	9.4%	6.7%	7.5%
	Total	**21.1%**	**18.3%**	**17.3%**	**24.5%**	**21.8%**	**22.7%**
Transportation	Deep Sea Freight Transportation	33.2%	28.8%	28.2%	31.6%	26.9%	27.0%
	Marine Passenger Transportation	35.1%	35.5%	36.4%	61.1%	62.8%	69.2%
	Marine Transportation Services	40.2%	30.3%	32.0%	39.7%	30.3%	32.2%
	Search and Navigation Equipment	6.7%	4.9%	4.0%	12.2%	10.1%	11.0%
	Warehousing	12.7%	9.5%	10.2%	12.9%	13.9%	15.5%
	Total	**20.4%**	**14.6%**	**24.6%**	**25.6%**	**19.6%**	**22.7%**
Ocean Economy	Total	**21.6%**	**17.8%**	**20.9%**	**25.8%**	**23.0%**	**27.4%**

Source: Calculated by authors and NOEP with data from the U.S. Bureau of Labor Statistics.

in employment and GSP for Ocean Economy sectors in the Gulf of Mexico and the United States is shown. Two things are noteworthy in this table. First, at the national level from 1990 to 2003, the only Ocean Economy sector to show any growth at all was Tourism & Recreation. Second, in the Gulf of Mexico, Tourism & Recreation is by far the fastest growing sector in terms of both employment and GSP. In fact, Tourism & Recreation GSP more than doubled in the Gulf of Mexico.

Besides putting Gulf Tourism & Recreation growth in context, the conclusion that the Gulf of Mexico is the U.S. Ocean Economy's industrial heartland is reinforced in table 4.9. Comparing the Gulf to the United States as a whole, GSP growth is faster for each sector except Transportation. In Living Resources, the decrease in the Gulf GSP is much lower than for the United States GSP. For employment, Gulf growth was higher than the national levels. Tourism's substantial increase has already been discussed, but for Construction and Ship & Boat Building, there was growth in the Gulf compared to decreases at the national level. In Transportation and Living Resources, there were decreases at both levels, but they were smaller in the Gulf Ocean Economy. Gulf employment fared worse than national employment in Minerals alone, where there was a 2% decrease, compared to no U.S. growth.

In the Ship & Boat Building sector, note that the Gulf region grew in employment over a period when there was substantial decrease in national employment. Within this sector, shipbuilding is the major industry. Both Mississippi and Louisiana are major shipbuilding states, making up more than a quarter of U.S. shipbuilding employment in 2003, which is a 10% increase since 1990.

U.S. and Gulf growth rates for employment and GSP for the individual industries within the Living Resources and Tourism & Recreation sectors are compared in table 4.10. The fishing industry in the Gulf lagged behind a weak U.S. sector, as employment decreases affected all sectors except for Fish Hatcheries & Aquaculture, where the Gulf trailed the country as a whole. In the Seafood Processing industry, which makes up the bulk of this NOEP sector, employment and GSP decreased less in the Gulf.

In the Tourism & Recreation sector, all Gulf of Mexico industries except Marinas, RV Parks & Campsites, and Zoos/Aquaria exceeded the employment

Table 4.9. *Growth Rates of Ocean Economy Sectors, 1990 to 2003*

Sector	Employment		Real GSP*	
	Gulf of Mexico	U.S.	Gulf of Mexico	U.S.
Construction	5%	−13%	34%	−8%
Living Resources	−30%	−42%	−14%	−52%
Minerals	−2%	0%	50%	43%
Ship & Boat Building	10%	−32%	18%	−6%
Tourism & Recreation	53%	32%	103%	48%
Transportation	−8%	−27%	51%	67%
Ocean Economy	29%	8%	61%	16%

* Deflated by GDP Implicit Price Deflator.
Source: Calculated by authors and NOEP with data from the U.S. Bureau of Labor Statistics.

Table 4.10. *Nominal Growth Rates for Selected Ocean Industries, 1990 to 2003*

	Employment		Real GSP*	
Industry	Gulf of Mexico	U.S.	Gulf of Mexico	U.S.
Living Resources	−30%	−42%	−14%	−52%
Fish Hatcheries & Aquaculture	25%	43%	136%	40%
Fishing	−84%	−56%	−114%	−89%
Seafood Markets	−62%	−65%	−61%	−82%
Seafood Processing	−13%	−30%	−12%	−16%
Tourism & Recreation	53%	32%	103%	48%
Amusement & Recreation Services	4%	−32%	35%	−56%
Boat Dealers	49%	18%	121%	69%
Eating & Drinking Places	64%	43%	89%	44%
Hotels & Lodging Places	23%	10%	118%	58%
Marinas	35%	38%	89%	60%
Recreational Vehicles Parks & Campsites	38%	48%	97%	113%
Scenic Water Tours	N/A	N/A	N/A	N/A
Sporting Goods	102%	27%	260%	82%
Zoos, Aquaria	389%	436%	855%	561%

* Deflated by GDP Implicit Price Deflator.
Source: Calculated by authors and NOEP with data from the U.S. Bureau of Labor Statistics.

growth rates of the United States, with the differences in the major sectors of Eating & Drinking Places and Hotels & Lodging Places especially notable. In GSP, the Gulf of Mexico industries exceeded the rates of growth in the comparable U.S. industries in every case except RV Parks & Campsites. This shows that the Tourism & Recreation growth in the Gulf was broadly based across essentially all industries involved.

In short, the Gulf of Mexico is seeing very fast growth in ocean-related Tourism & Recreation. This is seen not only in the direct measurement of Tourism & Recreation-related activity but also in the growth noted in Marine Passenger Transportation, which is primarily related to the cruise ship industry. But at the same time, the Gulf of Mexico is also increasing its role as the major producer of goods-related ocean activity within the U.S. Ocean Economy. These patterns suggest that, were they both to continue, there could be increasing conflict in the Gulf between the goods-related and the tourism and recreation-related industries over the inevitably scarce supply of prime coastal land. To see whether that is a possible concern, we need to look at the spatial distribution of the Ocean Economy within the Gulf of Mexico region at both the state and county levels.

The percentage of employment in each county in the Gulf Ocean Economy is shown in figure 4.14. The highest percentages, ranging from 20% to 30%, are

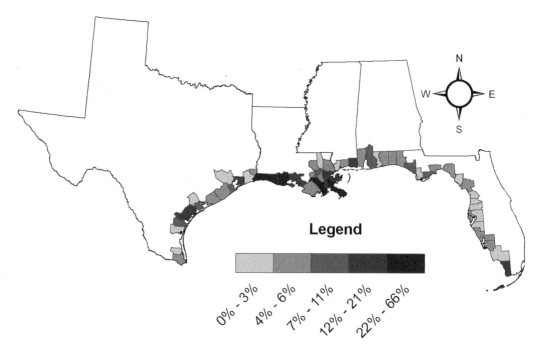

Figure 4.14. Percentage of county employment in ocean economy, 2003. Source: U.S. Bureau of Labor Statistics.

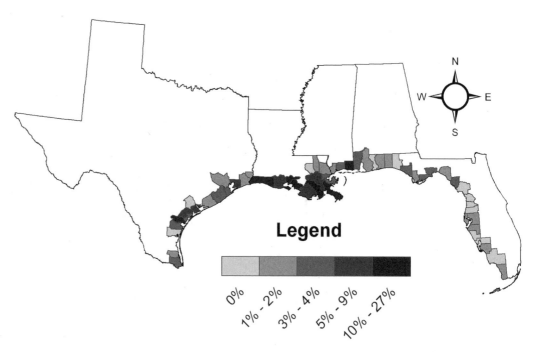

Figure 4.15. Proportion of county gross state product in ocean economy, 2003. Source: U.S. Bureau of Labor Statistics.

generally away from the cities in places like Monroe County, Florida; Plaque-mines Parish, Louisiana; and Jackson County, Mississippi. The lower proportions are in urbanized areas such as Tampa Bay and the Houston region. This is consistent with national trends, which show that most of the employment in the Ocean Economy is found in urban areas, but the Ocean Economy plays a larger role in more rural economies.

Besides Tourism & Recreation, all sectors of the Ocean Economy are goods-related. When these sectors are considered alone, the employment distribution picture changes somewhat. A concentration of Ocean Economy employment in the central Gulf of Mexico region becomes apparent, extending roughly from Jackson County, Mississippi, to Chambers County, Texas (fig. 4.15). To the east and west of this region, the Ocean Economy's role in local economies is shaped more by Tourism & Recreation.

Understanding the Non-Market Value of Coastal Recreation in the Gulf of Mexico

By the end of the twentieth century, more than 43% of all Americans participated in some form of marine recreation, according to the National Survey on Recreation and the Environment (NSRE), a nationwide study on coastal recreational uses (Leeworthy and Wiley, 2001). Americans flock to beaches and shores to swim, fish, boat, and view the natural scenery. The total number of people participating in all forms of marine recreation is expected to increase, with the largest increases expected for beach-going activities. By the year 2005, the number of Americans participating in coastal recreational activities was expected to have grown between 5% and 10% over 2000 levels, depending on the activity (Leeworthy et al., 2005).

Prior to 2005 and Hurricanes Katrina and Rita, the Gulf Coast was one of the nation's most important regions for coastal and marine recreation. The Gulf Coast shoreline stretches more than 1,600 miles, with Florida's 770 miles and Texas's 367 miles making up a full 70% of that total (table 4.11).

According to NSRE data, Florida and Texas rank among the nation's top five destinations for Americans who swim, fish, dive, and otherwise enjoy beaches, shores, and coastal wetlands. Florida's coastline, including its Gulf and Atlantic coasts, is the most popular in the nation, with nearly one in ten Americans visiting in 2000, which is more than 22 million visitors overall (Leeworthy and Wiley, 2001). Texas was fifth in the nation in 2000, with more than 6 million visitors. While having significantly smaller coastlines, the other Gulf states still attracted about 2 million visitors each in 2000. Alabama ranked sixteenth, Louisiana was twentieth, and Mississippi was twenty-second. The visitation rates and totals for all the coastal United States are summarized in table 4.12.

Coastal areas support a wide variety of recreational activities. In the Gulf of Mexico, visitors come to area beaches to swim, sunbathe, watch wildlife (espe-

Table 4.11. *Comparisons of Coastal and Estuarine Shoreline Lengths for the Gulf States*

	Alabama	Florida	Louisiana	Mississippi	Texas	Total
Coastal Shoreline Miles	53	770	397	44	367	1,631
% of Total State	3%	47%	24%	3%	23%	
Estuarine Shoreline Miles	607	5,095	7,721	359	3,359	17,141
% of Total State	4%	30%	45%	2%	20%	

All mileage numbers are rounded to the nearest 10 miles. Percentages are rounded and do not necessarily add up to 100%.
Source: NOAA, 1975.

Table 4.12. *Coastal Recreation Participation and Participants by State*

State	Participation Rate (% of national population)	Participants (where activities took place)	National Rank
Florida	**10.7**	**22,060,908**	**1**
California	8.71	17,654,215	2
South Carolina	3.14	6,469,023	3
New Jersey	3.02	6,224,769	4
Texas	**2.99**	**6,167,691**	**5**
North Carolina	2.7	5,576,629	6
New York	2.67	5,503,395	7
Massachusetts	2.38	4,904,006	8
Maryland	2.38	4,901,728	9
Virginia	2.37	4,878,313	10
Hawaii	2.2	4,540,543	11
Maine	1.82	3,753,337	12
Washington	1.66	3,429,729	13
Oregon	1.54	3,183,483	14
Rhode Island	1.28	2,641,812	15
Alabama	**1.24**	**2,549,078**	**16**
Connecticut	1.11	2,294,362	17
Georgia	1.1	2,262,763	18
Delaware	1.05	2,168,108	19
Louisiana	**1.05**	**2,165,830**	**20**
New Hampshire	1.03	2,120,282	21
Mississippi	**0.87**	**1,801,442**	**22**
Alaska	0.84	1,725,078	23
Washington, D.C.	0.13	258,559	24

Source: Leeworthy and Wiley, 2001.

cially birds), photograph scenery, boat, fish, and dive. The NSRE provides estimates for the number of participants and the number of days of participation for a variety of coastal activities in the Gulf states (table 4.13). It should be noted that the participation data for Florida include both the Atlantic and Gulf coasts. While site-specific studies may provide more accurate estimates of the number of users and participation days for specific locations, the NSRE provides the broadest range of activity estimates across the Gulf states and therefore forms the basis of our analysis. In the table, recreational fishing includes saltwater and mixed saltwater/freshwater areas in rivers and bays. Unless otherwise indicated, all activities take place on or in saltwater, including mixed fresh water and saltwater in tidal portions of rivers and bays. Beach going includes swimming, sunbathing, collecting seashells, walking, jogging, viewing birds or other wildlife, and any number of other activities that occur at the beach. This category is broad and should not be added to other more specific beach-related activities, to avoid double counting. Waterside activities include all

Table 4.13. *Recreational Participation by Activity and State Visited (in millions)*

Activities	Alabama		Florida (all coasts)		Louisiana		Mississippi		Texas		Total Days*
	Participants	Days	Participants	Days	Participants	Days	Participants	Days	Participants	Days	
Beach	1.249	11.842	15.246	177.153	0.629	4.042	1.042	8.679	3.851	35.239	237
Waterside	0.310	3.650	1.801	22.590	0.331	7.05	0.164	1.317	0.488	3.975	39
Swimming	1.022	8.203	14.033	161.098	0.398	4.59	0.563	6.739	3.076	29.59	210
Fishing	0.615	4.217	4.698	56.285	0.975	12.486	0.312	4.663	1.695	16.425	94
Motorboating	0.272	3.931	3.337	46.624	0.671	10.399	0.228	3.395	0.82	10.099	74
Personal Watercraft	0.139	0.669	1.626	14.540	0.136	**	0.07	**	0.272	2.906	18
Canoeing	0.019	n/a	0.019	n/a	0.019	n/a	0.01	**	0.046	**	
Kayaking	0.022	n/a	0.338	n/a	0.005	n/a	0.005	n/a	0.021	n/a	
Rowing	0.013	n/a	0.153	n/a	0.096	n/a	0	n/a	0.02	n/a	
Waterskiing	0.071	**	0.613	4.475	0.095	**	0.039	**	0.144	**	4
Bird Watching	0.351	4.719	3.373	77.952	0.387	9.114	0.317	7.248	0.805	16.051	115
Other Wildlife	0.364	6.435	2.846	50.264	0.385	10.555	0.235	2.381	0.745	12.604	82
Photography Scenery	0.441	7.369	3.920	96.591	0.596	16.902	0.427	8.856	1.193	32.188	162
Hunting	0.062	**	0.072	**	0.083	**	0.006	**	0.075	**	

*Rounded to the nearest million
**Too few to estimate
n/a Data not collected
Source: Leeworthy and Wiley, 2001.

activities already listed when conducted in waterside coastal areas that are not beaches.

Beach going dominates coastal recreational activities in the Gulf states, with just under 60 million visitor days to beaches in Texas, Louisiana, Alabama, and Mississippi during 2000. Forty-nine million of those visitors went swimming. During the same period, more than 177 million beach days and 161 million swimming days were enjoyed in Florida (including all coasts). Bird watching also is a popular activity, with more than 37 million visitor days involving bird watching and more than 65 million days involving photography in Texas, Louisiana, Alabama, and Mississippi. Similarly, nearly 78 million bird watching days and almost 97 million photography days were enjoyed in all of coastal Florida. Finally, fishing and boating draw many visitors to the Gulf Coast. In 2000, for Texas, Louisiana, Alabama, and Mississippi, anglers spent approximately 38 million visitor days and boaters almost 28 million. In all of coastal Florida there were over 56 million fishing days and nearly 47 million boating days in 2000.

Coastal and marine recreation generates value for participants, revenues for local businesses that support these activities, and taxes for a variety of levels of government. These activities generate both market and non-market impacts, which complicates the quantification of their overall economic impacts. The market impact of coastal recreation usually is assessed by examining how much money visitors contribute to the local economy through spending related to access, equipment, and goods and services such as ice and bait. Market-based studies commonly focus on gross expenditures, with fewer studies focusing on profits or taxes. While gross expenditures do not represent net benefits to the economy, they do capture the magnitude of importance that recreational expenditures have in the overall local economy. Spending by state residents represents a transfer of economic activity within the state. In other words, taxes generated by state residents are simply a transfer within the state from taxpayers to the treasury. Also, it is usually the case that spending by state residents would have taken place elsewhere in the state if not at the coast. Spending by out-of-state visitors, however, represents a direct economic influx for the state economy and is the base upon which additional tax revenues can be generated.

The non-market value of coastal recreation is more difficult to determine. Non-market values represent the value visitors place on the marine resources they use, beyond what they have to pay to access these resources. Non-market values often are associated with outdoor recreational resources, including recreational fishing sites, beaches, wildlife, and even views. The non-market values associated with coastal and marine resources have been shown to generate substantial economic value beyond the market expenditures generated by these resources. These non-market values represent the net economic value of the resource to the coastal visitor. While the literature recognizes non-market values that accrue to both users and non-users, we follow the policy of NOEP, which is to focus only on those non-market use values enjoyed by visitors to the coast as part of their use of the coast. Non-use values include options that people may be willing to pay for, such as one day using a resource; bequest values, which measure people's willingness to pay to help ensure that future generations are able to enjoy a resource; and existence values, which are sums people may be willing to pay simply to know that a resource exists. Use values tend to be estimated more frequently and with more precision than non-use values.

In the literature two categories of methods are used to estimate the non-market use value of coastal recreation—travel cost and contingent valuation

Table 4.14. *Non-Market Values for South Atlantic and Gulf Coast Beach Recreation*

Author	Location	Method	Asset	Consumer Surplus per person
Bell and Leeworthy (1986)	Florida	TC	Beach use	$19.43/day (average)
Bell and Leeworthy (1990)	Florida	TC	Saltwater beach use	$73.84/day
Bin et al. (2005)	North Carolina	TC, RUM	Beach use	$11.98–$84.49/ day (average)

Method: TC = Travel Cost Method; RUM = Random Utility Model.

Table 4.15. *Non-Market Values Associated with Wildlife Watching*

Author	Method	Location	Species	Consumer surplus per person day (in 2005 dollars)	Annual Non-Market Value*
Leeworthy and Bowker (1997)	Travel Cost Model	Florida Keys	Not identified	$108.35	$287 million
Colt (2001)	Unreported	Alaska	Sea birds	Min: $133 Max: $240	
Hall et al. (2002)	Contingent Valuation	California	Tide pools	$6.78/family visit	
Bosetti and Pearce (2003)	Contingent Valuation	England	Gray seals	For seeing seals in the wild: $14.5	
Johnston et al. (2002)	Travel Cost Method	New York	Not mentioned	$63.80	$35 million

* Values are rounded to the nearest $100,000.

methods. Travel cost methods are used to estimate the trade-offs visitors make between travel costs—including time and out-of-pocket expenses—and recreational opportunities.[6] Travel cost methods use real visitor behavior to estimate the non-market, or consumer surplus, value of coastal recreation. This is the value that coastal visitors place on a recreational trip beyond what they have to pay. But because the method requires considerable variation in the travel costs faced by visitors, it works best when applied to both residents and nonresident visitors from outside the immediate area. Authors also have used contingent methods to estimate values for coastal recreational use, especially when travel cost methods are difficult to apply. Contingent valuation methods are used to estimate both use values and non-use values and rely on surveys to elicit from visitors their willingness to pay to use, protect, or avoid damage to coastal recreational resources or access.

Below is a summary of studies providing estimates of non-market values that may be similar to those for coastal recreation in the U.S. Gulf of Mexico (tables 4.14, 4.15, 4.16a, and 4.16b). Because the goal of this examination is to provide values that may be similar to values for coastal recreation in the Gulf states, the review is limited when possible to studies of non-market values for coastal recreation in the Gulf of Mexico or southern United States. Note that the methods used to estimate these non-market values often differ among studies. In the following tables, which include brief explanations of basic methods,

Table 4.16a. *Non-Market Values for Atlantic and Gulf Coast Recreational Fishing Non-Residents*

Author	Location	Mode	Method	$/Day
Bell et al. (1982)	Florida	PC; P; S	CVM	61.86
Bockstael et al.* (1986)	South Carolina	P	CVM	97.92
McConnell et al.* (1993)	Mid-Atlantic/ Eastern States	PC; P; S	CVM	215.85
McConnell & Strand* (1994)	Delaware	PC; P; S	TC and RUM	17.07
	Delaware	PC; P; S	TC and RUM	18.51
	Florida	PC; P; S	TC and RUM	113.03
	Florida	PC; P; S	TC and RUM	135.86
	Georgia	PC; P; S	TC and RUM	66.06
	Georgia	PC; P; S	TC and RUM	70.12
	Maryland	PC; P; S	TC and RUM	44.67
	Maryland	PC; P; S	TC and RUM	45.73
	New Jersey	PC; P; S	TC and RUM	54.03
	New Jersey	PC; P; S	TC and RUM	56.95
	New York	PC; P; S	TC and RUM	96.35
	New York	PC; P; S	TC and RUM	98.31
	North Carolina	PC; P; S	TC and RUM	111.23
	North Carolina	PC; P; S	TC and RUM	114.81
	South Carolina	PC; P; S	TC and RUM	113.03
	South Carolina	PC; P; S	TC and RUM	114.44

Mode: PC = Party/Charter boat; P = Private boat; S = Shore.
Method: CVM = Contingent Valuation Method; TC = Travel Cost Method; RUM = Random Utility Model.
Mid-Atlantic/Eastern region includes the following states: DE, FL, GA, MD, NC, NJ, NY, SC, VA.
Source: As shown with * from Industrial Economics Incorporated.

Table 4.16b. *Non-market Values for Atlantic and Gulf Coast Recreational Fishing: Residents/Not Specified*

Author	Location	Mode type/Site Characteristics	Type of measurement and method	$/Trip	$/Day
Residents					
Bell et al. (1982)	Florida	PC; P; S	CVM		82.90
Downing and Ozuna (1996)	Texas	General boating	CVM	60.23–407.69 (mean of counties 171.11)	
Residential Status Not Specified					
Leeworthy (1990)	Florida	NS	TC	81.33	

Mode Type: PC = Party/Charter boat; P = Private boat; S = Shore.
Measurement: CVM = Contingent Valuation Method; TC = Travel Cost Method.
Florida: Includes Northwest Gulf, West Gulf, Northeast Gulf, Southwest Gulf, and Southeast Atlantic.

estimates provided have been converted to 2005 U.S. dollars, and figures are rounded when appropriate. Also, when possible, there is a breakdown of estimates based on the value per visitor per day. This makes it easier to compare results across studies and to understand how these values may compare to those generated by coastal recreation in the Gulf states.

Non-Market Value of Coastal and Marine Recreation   Coastal recreation generates direct economic benefits from visitors beyond the costs associated with getting to and using coastal resources. Changes in these non-market values, for better or for worse, reflect important changes in the net economic value of coastal resources and can result from changes in resource access, availability, or quality.

In this section, we review the literature to summarize estimates of the non-market values of coastal and marine recreational uses that are likely to be similar to those found in the Gulf states because original valuation estimates for the entire Gulf Coast do not exist. Generally, value estimates are available only for relatively isolated, local, or semi-regional examples. It is important to draw from this geographically well-specified literature to understand better the potential non-market economic value of coastal recreation at a more national level.

Benefits transfer is a term generally used to describe the practice of applying value estimates from the literature to new policy applications; for instance, using estimates from one area to estimate the values associated with another area. Benefits transfer techniques fall into two basic categories. The first is value estimates from the literature that are applied to the new policy site application by adjusting for differences in quality or accessibility between the new location and the previous location(s). The second category is the application of valuation models taken from other studies and applied directly to the policy site to generate new estimates of the potential economic value of that policy. In either case, the end result is usually a single number, or point estimate of value, given with some margin of error.

Generally, there is no clear agreement in the literature that the application of benefits transfer methods is accurate or appropriate (U.S. Environmental Protection Agency and Environment Canada, 2005). First, benefits transfer can only be as accurate as the original value transferred, and as already mentioned, determining the accuracy of any single study is difficult. Studies in the literature often fail to provide information about the details of the research methods, sample characteristics, and findings. Second, the personal characteristics of recreationists and the physical characteristics of the recreation sites at both the original site and the policy site are usually incompletely known. This makes attempts to adjust values or reapply models difficult.

Benefits Ranges   For this study, we apply a conservative approach to draw conclusions about the potential economic non-market value of coastal and estuarine recreation from the literature. First, we have attempted to provide a nearly exhaustive review of the peer-reviewed and, when possible, technical literature, on the estimated non-market value of recreational activities along the coasts and estuaries of the United States. The goal is to provide an overview of the range and magnitude of value estimates available. These valuation studies are organized by recreation type and region or state to illustrate how value estimates vary across uses and geography.

To extend understanding of these values further, we combine value estimates from the literature with regional and statewide estimates of user activity from

Table 4.17. *Estimated Non-Market Values for Selected Activities (rounded to the nearest $ million)*

Activities		Alabama	Louisiana	Mississippi	Texas	Florida (all coasts)
		Rounded to nearest million dollars				
Beach	high	$592	$202	$434	$1,762	$8,858
	low	$237	$81	$174	$705	$3,543
Swimming	high	$410	$230	$337	$1,480	$8,055
	low	$164	$92	$135	$592	$3,222
Bird Watching	high	$472	$911	$725	$1,605	$7,795
	low	$118	$228	$181	$401	$1,949
Other Wildlife	high	$644	$1,056	$238	$1,260	$5,026
	low	$161	$264	$60	$315	$1,257
Fishing	high	$422	$1,249	$466	$1,643	$5,629
	low	$253	$749	$280	$986	$3,377

Values cannot be added across activities due to potential for double counting.
Source: Estimated by authors based on cited studies in tables 4.12, 4.13, 4.14, 4.15, 4.16a, and 4.16b.

the National Survey on Recreation and the Environment (Leeworthy and Wiley, 2001). Like the value estimates, the NSRE figures are only estimates, and their accuracy varies from state to state depending upon the size of the population and the intensity of coastal activity. The NSRE data used also are limited in that they do not indicate the origin of visitors. Despite these limitations, the NSRE data provide the only consistent estimates of coastal recreation across the Gulf Coast.

Instead of answering the question, *What is the economic non-market value of coastal recreation?* we attempt to answer the question, *Is the economic non-market value of coastal recreation large and how does it potentially compare to the other economic activities along the coast described elsewhere in this chapter?* Throughout the rest of the chapter, value estimates are presented from the literature and user estimates from the NSRE at a regional level. The focus is on three activities for which the literature contains the most value estimates for the Gulf of Mexico and southern United States: beach going, wildlife viewing, and recreational fishing. Data presented include estimates of user activity from the NSRE combined with a range of both low and high estimates from the literature, which collectively provide a range for the potential non-market economic value of coastal recreational resources in the Gulf of Mexico, before Hurricanes Katrina and Rita. These aggregated values are summarized in table 4.17 and discussed in the sections that follow. The estimate ranges presented should be considered as *ballpark* figures that highlight the potential magnitude of non-market economic values. For each state, readers should use the data and literature presented to further explore and research the way in which these values are estimated.

Non-market values may differ between local visitors and non-local visitors. Unfortunately, the NSRE data on estimated participation do not reveal what proportion of visits are made by each category, although the raw data may contain this information. Non-market values also differ depending on the quality and nature of the coastal resources and proximity to population centers. As a result, the non-market value of an activity such as bird watching is likely to differ substantially across regions of the state. The literature on non-market estimates is not sufficiently rich to allow us to account for these differences easily. Therefore both a low and high estimate of the potential non-market value of recreational activities is offered, to account partially for the range of potential non-market values across users and regions.

Beaches   Warm waters and sandy beaches draw millions of visitors to the Gulf of Mexico, and in 2000 generated more than 267 million beach days at Gulf Coast beaches (including both Florida coasts); a beach day is defined as a visit by one person to a beach for one day. At least two studies (Bell and Leeworthy, 1986 and 1990) estimate the economic value of a beach day in Florida (adjusted to 2005 dollars) at between $19 and $74 (table 4.14). A more recent working paper (Bin et al., 2005) on beach use in North Carolina estimates the value of a beach day there at between $22 and $76 per beach day. These figures are similar to Pendleton and Kildow's (2006) estimated economic value for a beach day in California of $15 to $50 per beach day.

For this study, we use a range of $20 to $50 per Gulf Coast beach day to illustrate the potential non-market value of beach going in the Gulf states. The non-market value, in 2005 dollars, of beach use in Texas, Louisiana, Alabama, and Mississippi could have ranged from $1 billion to $3 billion on the basis of the NSRE's year 2000 visitation estimates of beach going. While it is impossible from these data to determine the proportion of Florida beach visits that occurred along the Gulf Coast, if one-third of Florida's beach days were along the Gulf, the total value would be between $1.2 billion and just under $3 billion.[7]

Bird Watching and Wildlife Viewing   Bird watching and wildlife viewing are also important contributors to the non-market value of coastal visitors along the U.S. Gulf of Mexico. The literature holds only a few examples of the non-market value of marine wildlife viewing (table 4.15), which range from tide-pooling in California, valued at less than $7 per family visit, to wildlife viewing in Alaska, with an estimated value of $133 to $240 per person per day for viewing seabirds. Leeworthy and Bowker (1997) estimated the economic value of general wildlife viewing in the Florida Keys to be $108 per person per day and $287 million annually for all visitors combined. To illustrate the potential value of wildlife viewing in the Gulf states before 2005, we use a lower bound of $25 and an upper bound of $100 per person per day. Using this range and the estimates of wildlife viewing provided by the NSRE, we estimate that the annual economic value of wildlife viewing in Texas, Louisiana, Alabama, and Mississippi would have been between just under $1 billion and about $3.7 billion annually, prior to Hurricanes Katrina and Rita. Again, Florida dominates the Gulf states in terms of total participation by wildlife viewers. If only half of the coastal bird and wildlife watchers in Florida were along the Gulf Coast, the non-market value of wildlife viewing along Florida's Gulf Coast would still range between just under $1 billion and $3.9 billion annually, which is almost the value of wildlife viewing in all other Gulf states combined.

Recreational Fishing   Recreational saltwater fishing also contributes significantly to the non-market value of coastal recreation in the Gulf states. In 2000, the NSRE estimates that more than 74 million person days were devoted to recreational fishing in the Gulf states. Most of these fishing days were in Florida. The literature on the non-market value of recreational fishing is extensive, with many examples from the Gulf states (tables 4.16a and 4.16b). For non-residents, the non-market value of a recreational fishing day in a Gulf state ranges from just over $60 (Bell et al., 1982) to more than $100 (McConnell et al., 1993). Both of these values are for fishing days in Florida. For residents, Downing and Ozuna (1996) estimated that the value for a fishing day in Texas ranged from $60 to more than $400. Bell et al. (1982) and Leeworthy (1990) both estimated values for a fishing day in Florida at just over $80 for residents and anglers of unspecified origin.

To illustrate the potential value of recreational fishing in the Gulf states before 2005, we use a lower bound of $60 and an upper bound of $100 per person day. Using this range and the NSRE estimates of recreational saltwater fishing, we estimate that the economic value of coastal and ocean recreational fishing in Texas, Louisiana, Alabama, and Mississippi could have been between $2.2 billion and $3.8 billion. Florida also generates numerous recreational angling opportunities. If only half of the recreational anglers in Florida fished along the Gulf Coast (note that in the National Ocean Economics Program data, more than 80% of Florida's saltwater marinas are in Gulf Coast counties), the non-market value of recreational fishing along Florida's Gulf Coast would still have ranged between just under $1.7 billion and $2.8 billion annually for the year 2000.

Conclusion

The non-market values associated with coastal recreation in the Gulf of Mexico generate billions of dollars in economic well-being for the nation. The Gulf Coast provides opportunities for Americans to boat, fish, hunt, swim, and view wildlife. These non-market values contribute directly to the quality of life of coastal visitors. As a result, damages to coastal resources result in a direct loss of these values that in turn represents a direct loss for the economic well-being of the region and the country.

The commercial importance of the Gulf of Mexico is fairly well understood. Oil and oil refining, transportation, tourism, and fishing all are important parts of the economic engine of the Gulf. However, far less is known about the market and non-market workings of the Gulf's coastal and marine recreational economy. We lack even a thorough baseline of coastal recreational activities within the Gulf region. Best estimates for most types of recreation, made by the National Survey on Recreation and the Environment (Leeworthy and Wiley, 2001) are limited in their ability to establish a baseline for coastal recreation in the Gulf. The NSRE lacks sufficient data for Louisiana, Alabama, and Mississippi to estimate adequately a baseline for many coastal recreational activities. The survey also does not discriminate between the Gulf Coast and the Atlantic and Caribbean coasts of Florida. As a result, we have only a partial understanding of the total level of participation in coastal recreation in the Gulf of Mexico.

Similarly, our understanding of non-market values for coastal recreation in the Gulf of Mexico is limited by the relatively few recent studies that have estimated environmental non-market values for Gulf Coast recreational activities. More studies of the non-market value of coastal recreation in the Gulf are needed. Further, to develop a good understanding of these values, it is important that studies be undertaken to value a wide range of activities and to do so across the entire Gulf.

For this study, we used the best available data from the federal government and the scholarly and gray literature to estimate the potential economic value of coastal recreation in the Gulf of Mexico prior to Hurricanes Katrina and Rita. Even using conservative estimates of values and visitation, we find that the non-market value of beach going, wildlife viewing, and fishing may each exceed $2 billion annually. In addition to these non-market values, coastal recreation generates substantial local revenues for coastal businesses. We do not even attempt to estimate these revenues here; but note that according to three recent papers prepared for the state of California, the volume of expenditures on coastal recreational activities is usually within one order of magnitude of the non-market value of these activities (Pendleton 2005; Pendleton and Rooke, 2005a and 2005b). Clearly, the potential magnitude of the economic value of coastal recreation in the Gulf of Mexico warrants a more comprehensive and consistent effort at data collection and research.

Notes

1. The NOEP includes market values based on the National Income and Product Account estimate from the U.S. Bureau of Census Bureau of Labor Statistics, data from the U.S. Department of Commerce Bureau of Economic Analysis, and the Quarterly Census of Economics of the U.S. Census Bureau. Natural resource production and values for living resources are from the U.S. Department of Commerce National Marine Fisheries Service of NOAA, and nonliving resources information is from the Marine Minerals Program of the U.S. Department of the Interior and state offices of the six states that produce offshore oil and gas.

2. The Bureau of Economic Analysis (BEA) defines gross state product (GSP) as the value added in production by the labor and property located in a state. GSP for a state is derived as the sum of the GSP originating in all industries in a state. In concept, an industry's GSP, referred to as its *value added,* is equivalent to its gross output (sales or receipts and other operating income, commodity taxes, and inventory change) minus its intermediate inputs (consumption of goods and services purchased from other U.S. industries or imported). Thus GSP is often considered the state counterpart of the nation's gross domestic product (GDP), the BEA's featured measure of U.S. output. In practice, GSP estimates are measured as the sum of the costs incurred and incomes earned in the production of GDP.

3. This estimate was computed by the authors using data from the U.S. Census Bureau, Bureau of Labor Statistics.

4. An establishment is a single reporting unit for employment and wages. These data were formerly known as the ES-202 data series. They are now reported as the Quarterly Census of Employment and Wages. They are compiled from reports filed by employers on a quarterly basis with each state's department of labor. State data files are transmitted to and maintained by the Bureau of Labor Statistics.

5. There is a small limestone, sand, and gravel industry in the Gulf of Mexico, but it is less than 0.5% of the value of the oil and gas industry. Because of disclosure issues (to avoid identifying individual firms), only the minerals sector is shown.

6. Travel cost methods include single and multiple site travel cost models, count data models, and a variety of site choice models, including random utility models.

7. Dr. Stephen Leatherman at Florida International University in Miami provides an annual ranking of the top ten beaches in the United States. In 2005, two of the nation's top ten beaches in the ranking were on Florida's Gulf Coast.

References

America's Wetland. 2008. http://www.americaswetland.com/.

Bell, F., and V. Leeworthy. 1986. *An Economic Analysis of the Importance of Saltwater Beaches in Florida.* Sea Grant Report 82. Florida Sea Grant College Program. Gainesville: University of Florida.

———. 1990. Recreational Demand by Tourists for Saltwater Beach Days. *Journal of Environmental Economics and Management* 22: 281–91.

Bell, F. W., P. E. Sorensen, and V. R. Leeworthy. 1982. *The Economic Impact and Valuation of Saltwater Recreational Fisheries in Florida.* Sea Grant Report 47. Florida Sea Grant College Program. Gainesville: University of Florida.

Bin, O., C. Landry, C. Ellis, and H. Vogelsong. 2005. Some Consumer Surplus Estimates for North Carolina Beaches. Working Paper, Department of Economics, East Carolina University.

Bockstael, N., A. Graefe, I. Strand, and L. Caldwell. 1986. *Economic Analysis of Artificial Reefs: A Pilot Study of Selected Valuation Methodologies.* Artificial Reef Development Center, Technical Report Series, vol. 6.

Bosetti, V., and D. Pearce. 2003. A Study of Environmental Conflict: The Economic Value of Grey Seals in Southwest England. *Biodiversity and Conservation* 12: 2361–92.

Coastal America. 2007. http://www.coastalamerica.gov/text/cwrp.html.

Colt, S. 2001. *The Economic Importance of Healthy Alaska Ecosystems.* Anchorage: Institute of Social and Economic Research, University of Alaska Anchorage. 6 pp.

Downing, M., and T. Ozuna, Jr. 1996. Testing the Reliability of the Benefit Function Transfer Approach. *Journal of Environmental Economics and Management,* 30: 316–22.

Florida Department of Environmental Protection. 2007. http://www.dep.state.fl.us/gulf/.

Hall, D. C., J. V. Hall, and S. N. Murray. 2002. Contingent Valuation of Marine Protected Areas: Southern California Rocky Intertidal Ecosystems. *Natural Resource Modeling* 15(3): 335–68.

Industrial Economics Incorporated. http://www.indecon.com/fish/default.asp.

Johnston, R. J., T. A. Grigalunas, J. J.Opaluch, M. Mazzota, and J. Diamantedes. 2002. Valuing Estuarine Resource Services Using Economic and Ecological Models: The Peconic Estuary System Study. *Coastal Management Journal* 30:47–65.

Leeworthy, V. R. 1990. An Economic Allocation of Fisheries Stocks between Recreational and Commercial Fishermen: The Case of King Mackerel. Ph.D. dissertation, Florida State University, Tallahassee.

Leeworthy, V. R., and J. M. Bowker. 1997. *Non-Market Economic User Values of the Florida Keys/Key West.* Linking the Economy and the Environment of Florida Keys/Florida Bay. National Oceanic and Atmospheric Administration, SEA Division, National Ocean Service, 41.

Leeworthy, V. R., J. M. Bowker, J. D. Hospital, and E. A. Stone. 2005. Projected Participation in Marine Recreation: 2005 and 2010. In *National Survey on Recreation and the Environment 2000.* Silver Spring, Md.: National Oceanic and Atmospheric Administration, National Ocean Service. 164 pp.

Leeworthy, V. R., and P.C. Wiley. 2001. Current Participation Patterns in Marine Recreation. In *National Survey on Recreation and the Environment 2000.* Silver Spring, Md.: National Oceanic and Atmospheric Administration, National Ocean Service.

McConnell, K., and I. E. Strand. 1994. The Economic Value of Mid and South-Atlantic Sportfishing. Department of Agriculture and Resource Economics. Information and values pertaining to this study were taken from www.indecon.com/fish/default.asp.

McConnell, K., Q. Weninger, and I. Strand. 1993. Testing the Validity of Contingent Valuation by Combining Referendum Responses with Observed Behavior. University of Maryland, Department of Agriculture and Resource Economics. Information and values pertaining to this study were taken from www.indecon.com/fish/default.asp.

National Marine Fisheries Service. http://www.nmfs.noaa.gov.

National Ocean Economics Program. 2007. http://noep.mbari.org/.

NOAA (National Oceanic and Atmospheric Administration). 1975. *The Coastline of the United States*. NOAA/PA 71046. Washington, D.C.: U.S. Department of Commerce, National Oceanic and Atmospheric Administration.

NOAA Coastal Services Center. 2008. http://www.csc.noaa.gov/crs/cwq/.

NOAA Fisheries Service. 2006. http://www.nwfsc.noaa.gov/research/divisions/fram/science.cfm.

———. 2008. http://www.nmfs.noaa.gov/.

North American Industry Classification System. http://www.census.gov/epcd/www/naics.html.

Pendleton, L. 2005. Understanding the Potential Economic Impact of Marine Wildlife Viewing and Whale Watching in California. California Marine Life Protection Act Initiative. Working paper.

Pendleton, L., and J. Rooke. 2005a. Understanding the Potential Economic Impact of Marine Recreational Fishing: California. California Marine Life Protection Act Initiative. Working paper.

———. 2005b. Understanding the Potential Economic Impact of SCUBA Diving and Snorkeling: California. California Marine Life Protection Act Initiative. Working paper.

Pendleton, L., and J. Kildow. 2006. The Non-Market Value of California's Beaches. *Shore and Beach* 74(2): 24–37.

U.S. Bureau of Labor Statistics. http://www.bls.gov/.

U.S. Census Bureau. http://www.census.gov/.

U.S. Commission on Ocean Policy. 2004. *An Ocean Blueprint for the 21st Century*. Washington, D.C.: U.S. Commission on Ocean Policy.

U.S. Department of the Interior, Minerals Management Service. http://www.mms.gov/.

U.S. Environmental Protection Agency and Environment Canada. 2005. Benefits transfer workshop proceedings. http://yosemite.epa.gov/EE/epa/eed.nsf/webpages/btworkshop.html.

5 Environmental Sustainability of Economic Trends in the Gulf of Mexico: What Is the Limit for Mexican Coastal Development?

ALEJANDRO YÁÑEZ-ARANCIBIA, JOSÉ J. RAMÍREZ-GORDILLO, JOHN W. DAY, AND DAVID W. YOSKOWITZ

Introduction

The increasing resource use and activity in Mexico's Gulf of Mexico coastal zone has created a series of environmental concerns that should be addressed in coastal management initiatives. The Mexican Gulf region, with the exception of the northern part of the state of Tamaulipas, is located in tropical latitudes (Yáñez-Arancibia and Day, 2004). Such areas are typically characterized by rapidly growing populations, deteriorating environmental quality, a loss of critical habitats, diminishing fish and shellfish populations, diminishing biodiversity, and increasing vulnerability to natural hazards (Cicin-Sain and Knecht, 1998; Westmacott, 2002). Poverty, institutional barriers, and resource depletion are commonly cited issues limiting advances in tropical coastal management efforts that might address these natural resource issues (Christie and White, 2000; Olsen and Christie, 2000). Nevertheless, there is an increasing capacity in tropical countries for coastal management with trends that include interdisciplinary research and management integration, development of new assessment and management techniques with special interest in ecosystem-level management, increased concern with climate change effects, and interest in using new eco-technologies for mitigating development impacts. Reliance on traditional knowledge and management systems is also growing, as is local participation in management efforts (Christie and White, 1997 and 2000; Windevoxhel et al., 1999; Yáñez-Arancibia, 1999 and 2000; Olsen and Christie, 2000; Day et al., 2003; Day et al., 2004; Yáñez-Arancibia and Day, 2004; Day et al., 2005; Day and Yáñez-Arancibia, 2008).

The major economic activities in the Mexican portion of the Gulf of Mexico include oil and gas extraction, power generation, lowland agriculture, fisheries, tourism, and port activities (Sánchez-Gil et al., 2004). The relationship between actual economic benefit and the expansion of these activities through infrastructure investments, agricultural expansion, increased fishing, and other efforts is not always clear. However, available data suggest a paradox whereby increased investments and activities are actually leading to reduced economic benefits, such as lower catch per unit fishing effort and lower worker incomes (Sánchez-Gil et al., 2004).

Such a lack of economic payoffs raises the question of whether inadequate environmental policies along with poor resource exploitation and planning can explain the region's limited economic growth. Clearly, the expansion of these economic activities has had severe environmental impacts on the Mexican coastal zone (Botello et al., 1996; Caso-Chávez et al., 2004) and has probably resulted in a loss of natural capital. Pollution and land-use change has modified the functional structure of ecosystems, diminished natural productivity, and decreased revenue derived from these economic activities, as has occurred in the north-central portion of the Gulf of Mexico (Day et al., 1997; Ko and Day, 2004), particularly in the Mississippi River and basin, and the oceanic dead zone.

Many economic and social issues related to resource use in the Gulf of Mexico are associated with the coastal zone, which extends from the coastal plain to the inner continental shelf. Other issues also involve the use of water resources, from coastal plain river basins to the estuarine plume, and are therefore water dependent. As a result, economic development in the coastal zone is extremely ecosystem dependent. At this time, however, it is not clear what the ecological limits may be on social and economic development, due in large part to a lack of available information about the coastal zone's carrying capacity.

In hopes of shedding light on what those ecological limits might be, as well as to increase understanding of the close ties between Mexico's economic health and the Gulf of Mexico's health, the goals of this study are to:

- Present a summary and description of the Gulf's ecological regions and major ecosystem processes related to key economic activities
- Integrate the environmental scenario and economic development relationships
- Present an overview of social and economic trends in the region
- Discuss the urgent necessity of economic, social, and environmental planning in the Mexican Gulf of Mexico coastal zone to achieve sustainable economic development in the region.

Defining Ecological Integrity   With ecosystems worldwide experiencing increasing stress, recognition of the need to safeguard the ecological processes on which all life depends has also increased. The United Nations World Commission on Environment and Development recognized that human society was dependent on sustainable use of the biosphere (Bruntland, 1987). But decisions about how to protect or rehabilitate biosphere resources are of necessity based on human values, and the intended goals for such actions are often not stated or unclear (Woodley, 1993).

A critical first step in planning for sustainable use of resources with clearly defined objectives is clearly defining the terms involved. The state of an ecosystem is commonly described in terms of ecosystem, biological, or ecological integrity, but these terms are often not easily defined. There are several published definitions in the literature, but most of them are subjective and general, with an ethical basis for understanding integrity. However, for the purposes of this chapter, we use a combined definition of these terms that incorporates both quantitative and ethical elements. *Biological integrity* is the capability of supporting and maintaining a balanced, integrated, adaptive community of organisms having a species composition and functional organization comparable to that of the region's natural habitat (Karr and Dudley, 1981). *Ecological integrity* describes ecosystem development that is optimized for its geographic location, taking into consideration energy input, available water, nutrient levels, and colonization history. The term implies that an ecosystem's structure and functions are unimpaired by human-caused stresses and that native species are present at variable population levels (Woodley, 1993).

Relationships between Ecological and Economic Regions

It is helpful to view the Gulf of Mexico coastal zone in terms of ecological subregions, previously defined for normative use by non-governmental organizations (NGOs), governmental organizations, and academics (fig. 5.1). Regional definitions are a management tool that aids in the definition of priorities for sustainable economic development as well as the assessment of conditions and trends for major ecosystems (Yáñez-Arancibia and Day, 2004).

Ecological Regions The Gulf of Mexico region includes three macrogeographical marine subregion*s*: the warm-temperate Gulf (WT), the tropical Gulf (TG), and the Caribbean coast of Mexico related to the Gulf (CCG). Only the northern part of the state of Tamaulipas lies in the warm-temperate region. This area is a transitional zone, as reflected in its seasonal temperature regime, which is influenced mainly by tropical currents in summer and a temperate climate during the winter. The bulk of the Mexican Gulf states, including Veracruz, Tabasco, Campeche, and Yucatán, are all within the tropical Gulf region, as is the southern tip of the Florida Peninsula. Quintana Roo, in the Mexican Caribbean along the eastern side of the Yucatán Peninsula, is within the CCG region. This area is the northern extent of the Mesoamerican Reef System, which is the second largest reef system in the world.

Each Gulf marine region can be viewed as a discrete system that results from the interaction of geology, geomorphology, oceanography, climate, freshwater drainage, physical and chemical features, coastal vegetation, wildlife, estuary shelf interaction, and human factors. Terrestrial ecological regions associated with these macro regions, such as the Mississippi Alluvial Plain and the Everglades, are also shown in figure 5.1, as are the relevant hydrologic units. Hydrologic units are geographic areas with drainage from a significant river basin through an important portion of the continent and extensive coastal plain. Examples in the Gulf include the Mississippi River basin and delta and the Usumacinta/Grijalva river basin and delta (fig. 5.1). Most of these units in the Gulf have high freshwater discharge into the coastal zone, and they often have important international impacts in the United States and Mexico (Yáñez-Arancibia and Day, 2004; Yáñez-Arancibia et al., 2006).

The economic importance and, to a large degree, the nature and biological composition of the Gulf of Mexico ecosystem are functions of its unique physical attributes. Some of the oceanographic features impacting the Gulf of Mexico, along with the main oil and gas fields and urban centers, are shown in figure 5.2. Fishing areas are also highlighted as the overlay of conch, demersal fish, lobster, and shrimp fisheries.

In general, marine waters enter the Gulf through the Yucatán Channel between Mexico and Cuba and exit through the Straits of Florida to become the Florida Current, and later the Gulf Stream, though actual currents are substantially more complex and vary on all time scales. Freshwater inputs are also significant in the Gulf. More than two-thirds of the United States' total freshwater drainage, and just under two-thirds of that of Mexico, drains into the Gulf (Yáñez-Arancibia and Day, 2004). A prominent oceanographic feature in the Gulf is the Loop Current. Large unstable rings of water detach from the Loop Current, bringing with them massive amounts of heat, salt, and water across the Gulf. Both the Loop Current and river discharge play important roles in shelf nutrient balance and coastal productivity in the Gulf of Mexico (Day et al., 1997; Lohrenz et al., 1999; Yáñez-Arancibia et al., 2003; Yáñez-Arancibia et al., 2004).

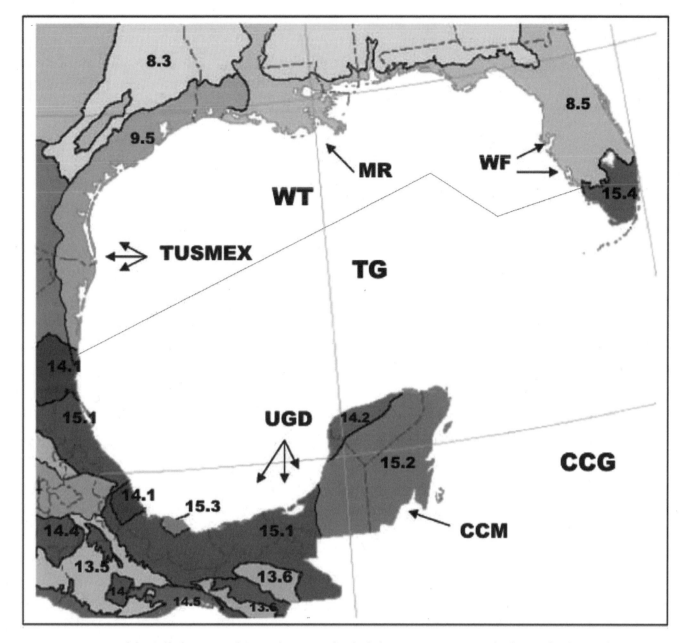

Figure 5.1. Diagram of the Gulf of Mexico indicating the geographic/hydrologic ecosystem units. The diagram has been redrawn to link terrestrial ecological regions and global marine regions as suggested by CEC/NAAEC/NAFTA (1997 and 2002, respectively), and it includes hydrologic units as suggested by Yáñez-Arancibia and Day (2004).

Terrestrial regions
8.5 = Mississippi Alluvial and Southeastern Coastal Plain
9.5 = Texas-Louisiana Coastal Plain
14.1 = Gulf of Mexico Dry Coastal Plains and Hills
14.2 = Northwestern Plain of the Yucatán Peninsula
15.1 = Gulf of Mexico Humid Coastal Plain and Hills
15.2 = Plain and Hills of the Yucatán Peninsula
15.3 = Sierra de los Tuxtlas
15.4 = Everglades
Global marine regions
WT = warm-temperate Gulf
TG = tropical Gulf
CCG = Caribbean coast of Mexico related to the Gulf
Hydrological units
WF = Western Florida rivers and groundwater discharge system
MR = Mississippi River basin and delta
TUSMEX = Texas estuaries and Laguna Madre U.S.-Mexico integrated by the Río Bravo delta
UGD = Usumacinta/Grijalva river basin and delta
CCM = Rio Hondo-Chetumal Bay on the Caribbean coast of Mexico

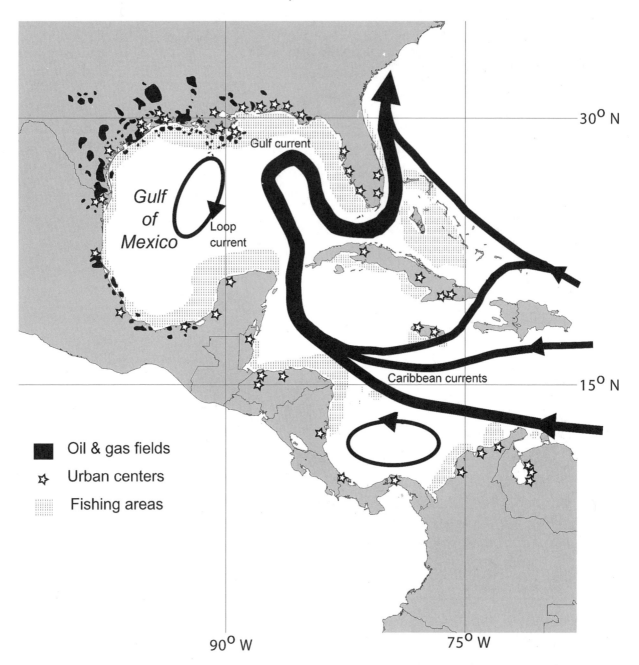

Figure 5.2. Distribution of oil and gas fields, urban centers with populations greater than 100,000, and fishing activities in the Gulf of Mexico region. The circulation links the ecological regions in figure 5.1 with populations of commercially important fish resources. Redrawn from Maul (1993).

Water-column dynamics on the broad, shallow Gulf continental shelves are strongly wind-driven out to depths of approximately 50 to 60 m and are extremely topographically diverse, with features such as smooth slopes, escarpments, knolls, basins, and submarine canyons. The Gulf of Mexico's southern coast is a wide, shallow carbonate platform that is widest at the eastern end of the Yucatán Shelf and narrows westward and northward along the Mexican coast.

Because of the semi-enclosed nature of the Gulf of Mexico and its general circulation pattern (fig. 5.2), human social and economic activities in one part of the Gulf can affect other areas. This was dramatically illustrated during the massive Ixtoc-1 oil spill in June of 1979 in the southern Gulf off Campeche, which affected the coastline well into Texas.

Hurricanes are a major climatic feature in the Gulf, causing perturbations to physical, biological, and human systems of the region. These impacts are currently magnified because of significant coastal wetland loss, sea-level rise, and dramatic population increases in the coastal zones in both the U.S. and Mexican areas of the Gulf.

Although several alternative methods have been considered to characterize ecological subregions in the Gulf of Mexico (USFWS, 1995; CEC/NAAEC/NAFTA, 1997, 2002; Frugé, 1999; Yáñez-Arancibia and Day, 2004), it is clear that no single ecological classification scheme provides an ideal framework for coupling environmental conditions with economic activities. However, the geographic units proposed by the U.S. Fish and Wildlife Service (USFWS, 1995), Frugé (1999), and the Commission for Environmental Cooperation, North American Agreement for Environmental Cooperation (CEC/NAAEC/NAFTA, 2002), and the hydrologic units proposed by Yáñez-Arancibia et al. (2003) and Yáñez-Arancibia and Day (2004), are advances in this framework process that can improve coastal management in the Gulf of Mexico using an ecosystem approach.

The ecosystem approach based on geographic and hydrologic ecosystem units applies well to the Gulf of Mexico because it focuses on coastal habitats where: (1) economic activities impact living resources, (2) resources are dependent on a gradient of areas from inland to Gulf coastal habitats, and (3) important ecological connections exist among the mainland drainage, estuaries, and coastal-marine areas on the continental shelf. This approach also addresses the Gulf indirectly in that coastal and nearshore habitats are affected by the water quality and quantity of streams entering coastal waters. For example, it recognizes a link between farming practices in the Midwest and water quality and fisheries productivity in the Gulf (Reyes et al., 1993; Mitsch et al., 2001). Erosion and nutrient runoff, natural or induced, result in sediment and nutrient transport downstream in the Mississippi and Usumacinta rivers (Day et al., 2003), producing significant impacts on the Gulf of Mexico. These inputs contribute to maintenance and accretion of coastal wetlands, which are vital as nursery areas for coastal ecological integrity, maintenance of biodiversity, and fisheries (Day et al., 2003). At the same time, excess nutrients that enter the system and are not trapped by marshes fuel algae blooms that ultimately cause anoxic bottom conditions known commonly as dead zones in U.S. and Mexican Gulf coastal waters (Rabalais et al., 1999; Yáñez-Arancibia et al., 2004).

Economic Trends   During the last decade, the proportion of the national population living in Mexico's six Gulf and Caribbean states remained relatively constant at about 15%. However, in the states of Quintana Roo (Caribbean coast), Yucatán, Campeche, and Tabasco (the southern Gulf), the growth rates were higher than the national average as a result of rapid growth in coastal municipalities, which are the equivalent of U.S. counties (table 5.1). Almost the entire populations of some states, such as Tabasco, Campeche, and Quintana Roo, live in coastal municipalities (Zárate Lomelí and Yáñez-Arancibia, 2003; León and Rodríguez, 2004; Sánchez-Gil et al., 2004).

In the Gulf region, the highest economic growth tends to be in the north and central states of Veracruz and Tamaulipas, while the highest population growth is in the south-central part of the region (table 5.1). This is a result of the dominant economic activities in each of these regions.

Table 5.1. *The Mexican States of the Gulf of Mexico and Caribbean Region: Population, Economic Trends, Resources Production, and Habitat Degradation*

	Coastal Municipalities	Total Municipalities	Total Inhabitants	Growth Rate	Agriculture (ha)	Fish Catch (tons)	Port Activities (% of national total)*
Tamaulipas	13	43	2,527,328	1.26	1,570,000	49,623 (15.4%)	5.8
Veracruz	45	210	6,737,324	0.94	1,600,000	135,745 (42.2%)	25.9
Tabasco	8	17	1,748,769	1.49	302,000	53,997 (16.8%)	8.8
Campeche	9	11	642,082	2.18	214,000	43,325 (13.5%)	17.5
Yucatán	25	106	1,556,622	2.03	787,000	35,409 (11.0%)	1.2
Quintana Roo	8	8	703,536	4.02	122,000	3,921 (1.2%)	3.5
Total	**108 (21%)**	**396 (100%)**	**14 mill. (100%)**	**1.7 (National mean)**	**4,595,000**	**322,020 (100%)**	**62.6**

	Coastal Inhabitants	Natural Gas (mill. ft³ year)	Crude Oil Production (mill. barrels per year)	Tourism Infrastructure (Hotel rooms)	Gross Domestic Product (% of national total)	Income %	Habitat degradation (% land-use change from total state original vegetation cover)	Habitat degradation (Main pollution pattern)
Tamaulipas	1,161,703	219,365	8.01	14,113 (15%)	3.2	20.5	60	Pesticides, metals
Veracruz	2,742,266	82,381	25.2	29,863 (31%)	4.5	51.8	87	Hydrocarbons, metals, pesticides, organic matter
Tabasco	870,628	483,611	189.6	4,459 (5%)	1.4	8.9	68	Hydrocarbons, metals, organic matter
Campeche	642,082	572,294	849.4	3,853 (4%)	1.2	4.2	39	Hydrocarbons
Yucatán	873,733	N/D	N/D	6,801 (7%)	1.3	9.2	83	Hydrocarbons
Quintana Roo	703,536	N/D	N/D	38,206 (39%)	1.3	4.9	30	Pesticides
Total	**7 mill. (50%)**	**1,358,298**	**1,072**	**97,295 (100%)**	**12.9%**	**100%**	**61.1 (Regional total)**	

* Data for 1999 only. Sánchez-Gil et al. (2004) calculate an average tonnage for imports and exports at 13 major Gulf of Mexico ports from 1990 to 1999 as 75% of the national total.
Sources: INEGI, 1999 and 2001; Zárate Lomelí et al., 1999 and 2004; CONAPO, 2000; INEGI/INE-SEMARNAT, 2000; PRC/INECOL, 2001; Ramírez-Gordillo, 2003; Zárate Lomelí and Yáñez-Arancibia, 2003; Sánchez-Gil et al., 2004; PEMEX-PEP, 2004a and 2004b.

The north-central portion of the area has more economic development and a more diversified economy. Important economic activities there include international trade—primarily non–oil and gas imports and exports through the Port of Veracruz and Altamira in Tamaulipas, fisheries, and the petrochemical and tourism industries. In general, most of these activities have stayed relatively stable over the last decade (INEGI, 2001; PRC/INECOL, 2001; Sánchez-Gil et al., 2004), with some relative growth in tourism and port development.

In the south-central portion of the region, the economy is less diversified. The main drivers there are oil production and transport through ports in the states of Tabasco and Campeche as well as tourism in Quintana Roo. Tourism in Quintana Roo is much higher than in other states (table 5.1), as reflected in hotel room availability, landscape, and ecosystem integrity. In general, these activities have increased in the last decade.

Primary economic activities for the region as a whole, including agriculture, cattle ranching, forestry, and fishing, have changed little over the past two decades (INEGI, 1999 and 2001; CONAPO, 2000; PRC/INECOL, 2001). Most of these activities are of a traditional character and concentrated in suburban areas. Even though a large proportion of the population is devoted to these activities, their impact on the national economy, at about 15%, is not proportional to the size of the workforce, which is approximately 68%. This is in strong contrast to technologically based industries with well-developed infrastructures, such as oil and gas, petrochemicals, and tourism, which contribute far more to Mexico's gross domestic product (INEGI/INE-SEMARNAT, 2000). To understand the Gulf economy better, it is useful to acquire a fuller understanding of the major industries in the region.

Oil and Gas   More than 80% of oil production and 90% of natural gas production in Mexico occurs in the Gulf of Mexico and its coastal plain. The Reforma-Tabasco fields and Campeche Sound are the most important oil regions of the country, and they are among the largest in the Western Hemisphere. In 2003 Mexico was the world's fifth largest oil producer, its ninth largest oil exporter, and the third largest supplier of oil to the United States, where over 70% of the total, or 1.8 million barrels a day, is exported (Pickering et al., 2003). The Gulf region alone produced 2 million barrels per day in 2003, with exports valued at $16.8 billion (PEMEX-PEP, 2004b).

Oil and gas revenues provide about one-third of all Mexican government revenues and will continue to be an important source of revenue for the immediate and long-term future. Yet Mexico increasingly relies on imports of natural gas from the United Status, with this figure reaching 397,086 million cubic feet in 2004, at an estimated total value of $2.3 billion.[1]

Petroleos Mexicanos (PEMEX), the state-owned oil and gas monopoly, has seen a steady increase in oil production over the last five years (fig. 5.3), averaging 3.38 million barrels per day in 2004, with natural gas production averaging 4,573 million cubic feet per day. This generated 561 billion pesos in total sales ($51 billion, with an 11:1 peso:dollar exchange rate) in 2004, from production alone (PEMEX-PEP, 2004a).

As energy prices rise due to increased demand worldwide, Mexico is looking to capitalize on its natural position, yet it finds itself in the unenviable position of having diminishing proven reserves. From the beginning of 1999 to the beginning of 2005, proven reserves have been cut in half (34 billion barrels of crude equivalent [bce] to 17 bce). Mexico's continuing industrialization, which requires

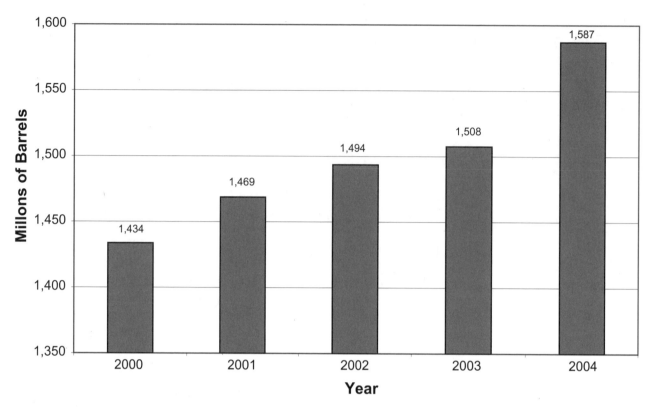

Figure 5.3. Total production of hydrocarbons. Source: PEMEX-PEP, 2005.

increasing amounts of energy, will require PEMEX to position itself better to feed the energy complex of the future. In August 2004 PEMEX announced that it had detected vast new oil deposits of about 54 billion bce in the Gulf that *could* double its total reserves to around 102 billion bce. This could potentially put PEMEX in a position to match production levels of Saudi Arabia or Russia at around 7.5 million barrels per day (Reuters, 2004). However, this new find has yet to materialize on PEMEX's reserve sheet as even being probable.

Fisheries   Throughout the 1990s the volume of fishery harvest from the Gulf of Mexico and the Caribbean varied between 200,000 and 350,000 tons per year. The greatest production levels are in the states of Veracruz, which brings in 43.6% of the total Gulf catch, Tabasco with 14.7%, and Campeche with 13.5% (table 5.1). The highest prices are in Tamaulipas and Yucatán, mainly for high-value commercial species such as lobster (*Panulirus argus, P. sygnus*), shrimp (*Farfantepenaeus duorarum, F. setiferus, F. aztecus*), octopus (*Octopus maya, O. vulgaris*), and conch (*Strombus gigas, S. costatus, S. alatus; Melongena corona*) (Sánchez-Gil and Yáñez-Arancibia, 1997; Sánchez-Gil et al., 2004).

In 2003 the Mexican Gulf states produced 283,153 tons of fish with an estimated value of 4.2 billion pesos ($381 million with an 11:1 peso:dollar exchange rate; CONAPESCA, 2003). The 1999 production of 322,000 tons was about 26% of the total national fishery harvest of 1.2 million tons. Such declines in capture yields will continue to put pressure on those who rely on this industry for their livelihood.

Port Activity   More than 75% of the tonnage of Mexican imports and exports, as well as leisure cruise traffic, occurs in thirteen important ports in the Gulf of Mexico. In 1999 the Gulf and Caribbean region accounted for 62% of the national shipping total. From 1995 to 1999 more than 80% of the annual total, including petroleum, was concentrated in the ports of Carmen, Campeche; Dos Bocas, Tabasco; Pajaritos, Veracruz; and Tuxpan, Veracruz.

The ports along the Gulf, excluding Campeche, handled more than 57 million tons of import and export cargo in 1999. This included the all-important shipments of petroleum. In 2004 this figure jumped to 75.5 million tons.[2] Even with a downturn in the global economy during this period, the ports of Altamira, Veracruz, and Dos Bocas continued to grow at a respectable rate.

Veracruz continues to be the key player in total tonnage for the region. It is not only the largest commercial port in the country but also the gateway for the automobile industry. In 2004 the port handled 70% of the total 687,000 automobiles shipped in and out of the country. In 2005 the Port of Veracruz expected to move 500,000 vehicles in and out of its facilities, which was 20,000 more than the previous year and a figure that warrants investment in the port. Currently, a highway is under construction in the region that will connect with other ports on the Gulf of Mexico, including Altamira and Tampico. That road could cut the costs that carmakers pay to ship out of Altamira, a trend that could take business away from Veracruz. Even though the Port of Altamira, located in the state of Tamaulipas, does not currently handle large numbers of automobiles, car shipments have grown significantly in the last two years. In 2004 the port exported 2,193 cars and imported 82,693 cars, compared to 629 in exports and 78,134 in imports in 2003 (Rueda, 2005).

Tourism Tourism in the coastal area is one of the most important sources of foreign income. In the Gulf and Caribbean, there are more than 1,900 hotels and other lodging accommodations, which is 23% of the national total. These facilities contribute about $600 million per year to the economy. In 2004 the number of rooms available was 93,301, excluding Tamaulipas.[3] Nearly ten years prior, the number of rooms available stood at 80,288, a figure which includes Tamaulipas (SECTUR, 2006). This significant increase in rooms and projected growth in the region will continue to fuel the regional economy.

The Mexican Gulf states, excluding Tamaulipas, welcomed 14.8 million visitors in 2004. Overwhelmingly the most popular destination was Quintana Roo, home to Cancun. Foreign visitors to this city outnumbered Mexican visitors by more than 4 to 1. By comparison, in Veracruz, the second most visited state, the opposite was the case, with Mexican nationals outnumbering foreign visitors 23 to 1.

Environmental Scenario and Economic Development Relationships Mexico's Gulf coastal zone habitats include freshwater bodies of the coastal plain, freshwater wetlands, lower river basins, coastal dunes, coastal lagoons and estuaries, salt marshes and mangroves, and the adjacent sea. The coastal zone has highly diverse physical and ecological processes, has a great richness of species and natural resources, and forms a very important regional system from ecological, economic, and social-cultural points of view (Kumpf et al., 1999; Caso-Chávez et al., 2004).

However, many human activities in the coastal zone affect this regional ecosystem. These include extraction of natural resources through fisheries in lagoons and estuaries, forestry, oil and gas extraction, agricultural activities, urban development, shipping, and tourism. All are conducted without sufficient strategic environmental planning because there is currently no program for integrated coastal management in the Mexican portion of the Gulf of Mexico (Reyes et al., 1993; Zárate Lomelí et al., 1999; León and Rodríguez, 2004; Sánchez-Gil et al., 2004; Zárate Lomelí et al., 2004; Zárate Lomelí and Yáñez-Arancibia, 2004). Consequently, these activities often lead to contamination and habitat destruction (table 5.1), inducing uncertainty in economic development and leading to conflicts of interest with environmental values.

Today Mexico has a total of 6,500 km² (2,500 mi²) of inland wetlands, mainly associated with lakes, lagoons, and rivers (Yáñez-Arancibia et al., 2006). The country's main wetlands, however, lie on the coast, covering an area of some 12,500 km² (4,800 mi²). Many of Mexico's wetland species are rare or endangered, notably aquatic vegetation, freshwater and marine turtles, manatees, caiman, crocodiles, a number of fish species, and many resident and migratory birds.

The major threats to Mexican wetlands are industrial development and agriculture. The cultivated area in the six coastal states of the region in 2000 was 217,250 km² (83,880 mi²), with an annual production of 4,227,923 tons, mainly corn, beans, wheat, rice, soybeans, sugarcane, cotton, and sorghum. Yet coastal lagoons and estuaries, which cover 16,000 km² (6,000 mi²), including water surface area and surrounding wetlands, are capable of producing food equivalent to that produced on 160,000 km² (62,000 mi²) of Mexico's agricultural land. Lagoons—such as Tamiahua, Tamaulipas; Alvarado, Veracruz; Terminos, Campeche; and the wetlands in Tabasco—have the highest productivity of any habitat type in Mexico (Yáñez-Arancibia et al., 2004; Yáñez-Arancibia et al., 2006). If adequately managed, they could, in a year, produce close to 180 kilograms of oysters per hectare (160 pounds per acre), which is more than ten times the amount of beef produced on drained wetlands.

Nonetheless, wetland loss continues at an alarming rate. Land-use changes in each one of the coastal states are mainly responsible for deterioration of wetlands and other natural areas (table 5.1). From the original vegetation cover, deterioration has been estimated as 60% in Tamaulipas, 87% in Veracruz, 68% in Tabasco, 39% in Campeche, 83% in Yucatán, and 30% in Quintana Roo. This means that on average, a dramatic 61% of the total original vegetation cover in the Gulf of Mexico coastal zone states has been lost (Zárate Lomelí and Yáñez-Arancibia, 2004).

The Mexican coastal zone in the Gulf, including the portion in the Caribbean region, forms a highly productive ecosystem that constitutes nearly 30% of the total national Mexican coast. The Gulf coastal area contains more than 65% of the coastal plain forest reserves and 50% of shrimp fisheries in Mexico. Of the five most important industrial ports in the country, three are located in this region: Altamira, Veracruz, and Coatzacoalcos.

Production in the coastal area of the Mexican Gulf and Caribbean was no more than 13% of the national GDP in the year 2000 (table 5.1; Sánchez-Gil et al., 2004). Within the region, the north-central states of Tamaulipas and Veracruz contributed more than 50% of regional production. The state of Veracruz has the largest economy of the region, contributing about 4.5% to 5.5% of Mexico's total GDP. In 2000 Tamaulipas contributed 3.2%, Veracruz 4.5%, Tabasco 1.4%, Campeche 1.2%, Yucatán 1.3%, and Quintana Roo 1.3% (table 5.1). Ramírez-Gordillo (2003) and Sánchez-Gil et al. (2004) have pointed out that the Gulf region's low contribution to GDP, and the decrease in some productive economic activities, could be partially explained by the decrease in

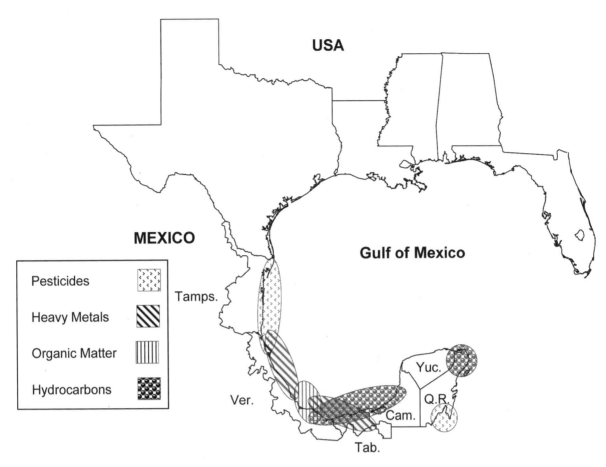

Figure 5.4. Pattern of distribution of major pollution and habitat degradation in the Mexican part of the Gulf of Mexico (Tamps = Tamaulipas, Ver = Veracruz, Tab = Tabasco, Cam = Campeche, Yuc = Yucatán, Q.R. = Quintana Roo). This diagram can be coupled with ecological regions in figure 5.1, urban centers in figure 5.2, and economic trends in table 5.1. Based on Botello et al. (1996) and Caso-Chávez et al. (2004).

ecosystem integrity, particularly the dramatic level of land-use change and pollution (table 5.1, fig. 5.4). In other words, diminishing environmental quality could be threatening the sustainability of economic activities in the Mexican portion of the Gulf of Mexico.

Discussion

Habitat Degradation, Economic Development, and Environmental Sustainability
Numerous conflicts exist between development interests and the local economies of the Mexican Gulf states regarding coastal resources (Zárate Lomelí et al., 1999; Day et al., 2003; Zárate Lomelí et al., 2004). The intensity of exploitation of these resources is causing serious environmental and habitat deterioration (fig. 5.4). This is a serious concern because there is a relationship among ecological regions (fig. 5.1), major urban centers (table 5.1, fig. 5.2), economic trends (table 5.1, fig. 5.3), and major pollution patterns in the Mexican part of the Gulf of Mexico (table 5.1, fig. 5.4). The link between economic activities and habitat degradation is clear (table 5.1), and this problem is magnified by the lack of environmental strategic planning for the coastal zone (Zárate Lomelí et al., 1999; Zárate Lomelí et al., 2004).

The Mexican Gulf region's main pollution patterns (fig 5.4) reflect major economic trends in each one of the Mexican Gulf states (table 5.1). This connection stems from a combination of forces, including cultural processes and natural ecosystem changes as well as industrial activities, land-use change, and habitat alteration and destruction. For instance, Tamaulipas receives considerable freshwater discharge from both the United States and Mexico from rivers including the Rio Grande/Río Bravo and the Tamesi/Pánuco river and basin, with a combined contribution of 480 m^3/s during the period from 1973 to 2000. All these rivers are clearly characterized by pesticide and metal pollutants (fig. 5.4).

Veracruz is being highly degraded by port activities, hydrocarbons and petrochemical industries, and agriculture and cattle ranching, all of which reflect the highest land-use change (87%) in the region. Urban development, cattle ranches, and sugarcane production and processing are all significant contributors of organic matter contamination to the area via freshwater rivers (table 5.1, figs. 5.3 and 5.4). Major discharges include the Pánuco River with a flow of 459 m^3/s (1972 to 2000), La Antigua River with 57 m^3/s (1951 to 2000), the Jamapa River, Río Blanco, and Papaloapan delta with 600 m^3/s collectively (1973 to 2000), and the Coatzacoalcos River with 450 m^3/s (1953 to 2000).

The Tabasco and Campeche coastal zone is strongly influenced by the Grijalva/Usumacinta river basin and delta, which has the second highest river discharge in the Gulf of Mexico, after the Mississippi River. This is one of the largest deltas in North America with an area of 20,000 km^2 integrating the rivers Usumacinta, Grijalva, and Mezcalapa, and a watershed basin area of 177,987 ha in Campeche, 724,547 ha in Tabasco, 2,175,718 ha in Chiapas, and 4,241,271 ha in Guatemala (Day et al., 2003). Thus these rivers flow to the coast after draining a tremendous continental surface area mainly dedicated to tropical agriculture and cattle ranching. The combined discharge of the Grijalva and Usumacinta rivers has been estimated between 3,723 and 4,402 m^3/s (Day et al., 2003). This is likely a partial explanation for why Tabasco and Campeche have significant problems with organic matter, pesticide, and nutrient pollution in the coastal zone (fig. 5.4).

Campeche has the highest oil and gas production of the region, which is reflected in the highest hydrocarbon pollution in the region's coastal zone (table 5.1, figs. 5.3 and 5.4). Yucatán has a large agricultural area of 787,000 ha, which has caused 83% land-use change (table 5.1). The state also has important hydrocarbon pollution, likely due to shipping throughout the Yucatán Channel (figs. 5.3 and 5.4). Finally, Quintana Roo has the most tourism in the region and the highest growth rate but lower land-use change. As in Yucatán, there is significant hydrocarbon pollution in the Quintana Roo coastal zone, partially as a result of shipping throughout the Yucatán Channel and the Caribbean coast (Gold, 2004).

Loss of wetlands and other coastal habitats is a problem that characterizes the Gulf of Mexico coastal region in both Mexico (table 5.1) and the United States (Harwell, 1998; Day et al., 2003; Day et al., 2004a; Ko and Day, 2004). This has a tremendous impact on coastal ecosystem integrity (Day et al., 1997; Rabalais et al., 1999; Mitsch et al., 2001; Caso-Chávez et al., 2004; Day et al., 2004b; Yáñez-Arancibia et al., 2004). As a result, the main sectors of the economy affected are the primary ones. The predominant productive activities are agriculture, cattle ranching, forest exploitation, and fishing (Zárate Lomelí et al., 1999; Sánchez-Gil et al., 2004; Zárate Lomelí et al., 2004).

These rural and traditional activities often suffer with the development of modern high-technology industries, such as oil and gas exploitation and petrochemical activities. For example, over the past two decades, the deltaic region of Tabasco-Campeche and the coastal plain of the state of Veracruz have undergone remarkable economic transition (León and Rodríguez, 2004; Sánchez-Gil et al., 2004). Starting from isolated, mostly rural regions, these areas have grown through a number of important transformations. These have included large-scale commercial cattle ranching; the establishment of tropical farming plantations; the exploration, exploitation, transport, and refining of oil and gas; the development of petrochemical industries; and urban expansion due to both industrial activities and tourism development. These transitions have altered the landscape of the coastal area dramatically.

Besides the pollution problems previously noted, it is clear that the oil and gas industry in the southeastern portion of the Mexican Gulf coast has had a dramatic impact in the coastal zone (Botello et al., 1996; Day et al., 2004c; Gold, 2004; Ko and Day, 2004; Reyes et al., 2004). The exploration, exploitation, and transport of hydrocarbons have led to various environmental problems, including water, air, and soil contamination and wetland loss (Botello et al., 1996; Day et al., 2003; Herrera-Silveira et al., 2004; Day et al., 2005). The adoption of preventive measures seems to be the best alternative for mitigation. Solutions to these problems will be more effective when there are studies of economic valuation of natural resources and environmental impacts. These should be routinely incorporated into analyses of economic sustainable development in the region's coastal zone.

At present, the economic development pressure in the Mexican Gulf coastal zone is arriving at a threshold, beyond which is the collapse of environmental sustainability. Sánchez-Gil et al. (2004) have clearly shown that natural resources under exploitation in the coastal zone are coastal Gulf-dependent. Many economic and social issues are geographic in nature and associated with the coastal zone from the coastal plain to the inner continental shelf. These issues are also related to the use of water resources from lower river basins to the estuarine plume and are therefore water-dependent (Sánchez-Gil and Yáñez-Arancibia, 1997; Sánchez-Gil et al., 2004). In other words, the economic development in the Gulf of Mexico coastal zone is ecosystem-dependent and strongly based on ecological integrity, which is at risk because of land-use change, pollution, and urban and industrial expansion.

Global climate change is an environmental variable often ignored or undervalued in Mexican coastal development actions (Yáñez-Arancibia and Day, 2005). In Mexico, global climate change could lead to increasing temperature, changing precipitation patterns and producing heavy rains, floods, and drought, increasing intensity and frequency of tropical storms and hurricanes, and sea-level rise (Emanuel, 2005; Yáñez-Arancibia and Day, 2005; Ortiz Pérez et al., 2008). Similar impacts may be seen in the U.S. littoral zone in the Gulf (Day and Templet, 1989; Twilley et al., 2001, LeRoy Poff et al., 2002; Scavia et al., 2002; Ning et al., 2003; Day et al., 2005).

Such climate change effects could have tremendous human, economic, and ecological impacts in the near future. There is a broad consensus in the scientific community that human activity is affecting global climate (IPCC, 2001). Global warming will lead to accelerated eustatic sea-level rise of as much as 40 to 80 cm by the end of the twenty-first century in the Gulf of Mexico. This increase in sea level must be added to subsidence to obtain the relative sea-level rise that coastal wetlands in the Mississippi delta in the United States, and Usumacinta/Grijalva delta in Mexico, will be subject to during the twenty-first century. The sea-level rise in those deltas is predicted to increase from about 1 cm/yr to 1.3 to 1.7 cm/yr within this century, a 30% to 70% increase (Day et al., 2005; Ortiz-Pérez et al., 2008). Accelerated sea-level rise, increased frequency of tropical storms and hurricanes, and higher river discharges will lead to a significant increase in flooding duration in the Mexican portion of the Gulf of Mexico. Unless wetlands can accrete vertically at the same rate as water-level rise, coastal vegetation will become progressively more stressed and will ultimately die (Day et al., 1997). This is a critical concern to coastal states from southern Veracruz to Tabasco, and in Campeche as well.

Another factor that is likely to affect environmental management in coming decades is the availability and cost of energy. Over the past decade, increasing information suggests that world oil production will peak within a decade or two, implying that demand will consistently be greater than supply and that the cost of energy will increase significantly in coming decades. This information has come primarily from petroleum geologists with long experience in petroleum production (Campbell and Laherrère, 1998; Deffeyes 2001 and 2002; Bentley, 2002; Hall et al., 2003). If the end of the era of cheap energy is indeed near, energy-intensive management methods will become increasingly expensive and untenable. Sustainable management of the southern Gulf will have to consider less energy-intensive, less expensive options for restoring deteriorating coastal marshes in the post–oil peak era. This approach of using the energies of nature to the greatest extent possible is called ecological engineering, and it involves using small amounts of fossil fuel energy to channel much larger flows of natural energy (Odum, 1971; Mitsch and Jørgensen, 2003). Ecological engineering offers both a conceptual and a practicable approach for long-term management of deltas in an era when fossil fuel energy will become much more expensive. An example of ecological engineering is the use of treated sewage effluent to enhance and restore coastal wetlands. Not only is the approach of wetland assimilation more economical and energy efficient, it also enhances the ability of coastal wetlands to survive sea-level rise (Day et al., 2003; Day et al., 2004a; Day et al., 2005).

Conclusion

Many economic development activities have had an impact on wetland loss in the coastal zone of the Mexican Gulf. These include (1) urban expansion and social development, (2) the development of oil, gas, and petrochemical industries, (3) expansion of agricultural and cattle ranching activities, (4) expansion of tourism and road infrastructure, and (5) uncertainty about the compatibility of economic development and environmental sustainability.

For the environmental sustainability of economic trends in the Mexican Gulf coast, the most important word should be *planning*. At present, an integrated management plan for the coastal zone in the Gulf of Mexico does not exist, and because of this there are no coherent environmental planning activities involving the three levels of governance (federal, state, and municipal) in Mexico. Sustainability of social and economic development must be based on the functioning of natural ecosystems, integrating natural resource exploitation with natural ecosystem dynamics, ensuring equity in social and economic development, and preserving ecosystem integrity. The continuing contrasting relationships among abundant natural resources, dramatic habitat degradation, high levels of pollution, and a low contribution from the region to GDP reflects the lack of coastal environmental strategic planning in the Mexican Gulf states. Even more, climate change effects must be incorporated in every project, every design, and every engineering action because coastal erosion, saltwater intrusion, sea-level rise, and changes in precipitation are happening now and must be accounted for in any development initiative. Any project started today and intended to be producing results ten years from now will be affected both by the absence of environmental strategic planning and by climate change effects.

In coastal and deltaic settings with high seasonal river input, the ecosystem is developed and maintained by a series of energetic pulses that occur on different temporal and spatial scales (Day et al., 1997). These include the switching of river channels every few hundred years, large storms and river floods, annual river floods, winter frontal systems, and tides. The maintenance of these pulses is critical to maintaining healthy ecosystems in an area with a high level of economic development. This approach can be used to integrate environmental functions with economic and ecological processes, which will allow better environmental and development decisions to be made and will lead to more sustainable development.

Another important criterion to consider in environmental planning is the intrinsic economic value of coastal resources. These resources represent a *natural capital* that supports the economic health of society. The goods and services provided by the natural capital represent the *interest* generated by human investment in natural ecosystems (Costanza et al., 1989; Costanza et al., 1997; Yáñez-Arancibia, 1999). This is the reason that healthy ecosystems support a healthy economy, as illustrated in fig. 5.5. The Gulf of Mexico is rich in natural capital that provides a source of income for many, a prime recreational resource, a nurturing ground for rich and varied aquatic life, and an indispensable food supply for a rapidly growing global population. In fact, there are so many values associated with the Gulf that it is difficult to produce an inclusive list.

However, this valued resource has been taken for granted by many in Mexico. But the country has passed the point where we can continue to ignore the warning signs. There are enough documented threats to the ecosystem that we simply cannot continue to take the Gulf's future health and high level of productivity for granted. If pathway 1 in fig. 5.5 is to be taken, current approaches to development must be changed. Mexico needs to be proactive with sustainable development. Natural and human activities or inactivity, both in the Gulf and in adjacent areas, have adversely affected the functioning and value of the Gulf as a unit, and this will likely get worse unless action is taken. Impacts may be immediate, a decade or more in coming, or a combination thereof. Action is called for. We must be sober and objective in our appraisal of threats. We can solve some of the problems facing us before they become acute and require emergency action. Yet time is of the essence; it is in no one's best interest to let small problems become big ones. Many of the problems of the Gulf ecosystems need priority attention, and creating a model management process with attention given to equity issues is necessary (Day and Yáñez-Arancibia, 2008).

Some of the problems affecting the Gulf can be solved by technological means, but technology is by nature data intensive and will require expanded and interactive databases on existing Gulf conditions as a condition for effective action (Day et al., 2004a; Day et al., 2004b). And many of the problems of the Gulf cannot be solved by technology but demand a change in the way of doing things. The Gulf of Mexico is a source of natural resource *capital*, actually a living *bank account,* which can be depleted and must therefore be managed with long-term objectives in mind, lest the Gulf fail to meet our future needs; and keeping its ecological integrity is a key component in the equation (fig. 5.5).

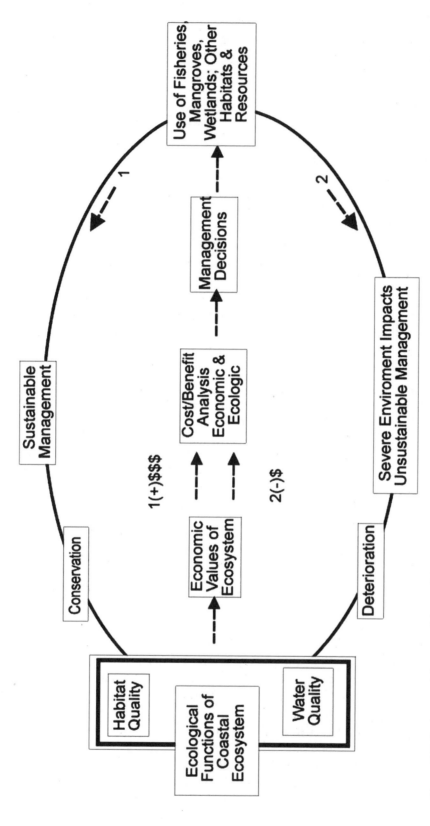

Figure 5.5. Environmental sustainability and the economic development of the coastal zone, indicating the importance of ecosystem feedback to sustainability; the economic values of ecosystems and resources; support for management decisions; and sustainable development perspective as a function of ecological integrity measures such as habitat quality, water quality, biodiversity, and ecological functions. Modified from Yáñez-Arancibia (1999).

In valuation terms, the language of currency is universal, the *coin of the realm*, as it were. But there are persons who do not accept the traditional Western valuation process, which quantifies and then assigns monetary values to resources. Alternate methods of valuation range from the sophisticated *unit of habitat value*, used by some agencies, to assessing a spiritual consideration of people's right to use Earth's resources temporarily assigned to their care (Delta Assembly, 1992). Such approaches can help fill voids in traditional valuation approaches. However, currencies such as the dollar or the peso are a widely accepted valuation standard, and we acknowledge the need for monetary valuation as a necessary component of a public education campaign about Gulf of Mexico extrinsic and intrinsic values. For example, a multitiered valuation structure can consider many components, such as ecotourism value, minerals and petrochemicals value, wildlife value, commercial value, aesthetic value, climate value, fisheries value, biological value, and social or cultural value. We conclude that what is in jeopardy is the energy required to *run* the vast renewable ecological system of the Gulf that produces so many resources (table 5.1); creates essential habitat as a life support system, both temporary and permanent; and provides an economically, socially, and aesthetically significant recreational base.

Unfortunately, many believe there is no limit to economic growth in the Gulf of Mexico. In order to develop an alternative, sustainable view of the Gulf, we need to develop and apply methods of analysis of environmental and economic information and to integrate this information into economic and ecological plans for each one of the coastal states of the Gulf and Caribbean region in Mexico. This will allow the limits of economic development to be defined by threats to the ecological integrity of ecosystems. In this approach, we can see that these are limits that humans should respect in order to be able to understand and quantitatively measure the environmental sustainability of the socioeconomic development of the Mexican coast of the Gulf of Mexico.

Notes

1. Author's own calculations, and Energy Information Agency, *U.S. Natural Gas Exports by Country,* http://tonto.eia.doe.gov/dnav/ng/ng_move_expc _s1_a.htm.
2. The 2004 figures for Coatzacalcos were extrapolated from February 2004 figures. All these values were gathered from the individual ports.
3. The figure does not include the state of Tamaulipas due to insufficient data.

Acknowledgments

This study is the result of cooperative activities between the Instituto de Ecología A.C., México, Projects INECOL 902-19-752 and Centla 91-41; the Department of Oceanography and Coastal Sciences, Louisiana State University, NOAA Grant no. NA16G2249; and the Department of Finance, Economics and Decision Sciences, College of Business and Harte Research Institute for Gulf of Mexico Studies, Texas A&M University–Corpus Christi.

References

Bentley, R. W. 2002. Global Oil and Gas Depletion: An Overview. *Energy Policy* 30: 189–205.

Botello, A. V., J. L. Rojas, J. A. Benítez, and D. Zárate Lomelí (eds.). 1996. *Golfo de México contaminación e impacto ambiental: Diagnóstico y tendencias.* EPOMEX Serie Científica. Campeche, México: Universidad Autónoma de Campeche. 650 pp.

Bruntland, G. 1987. *Our Common Future.* New York: Oxford University Press. 400 pp.

Campbell, C. J., and J. H. Laherrère. 1998. The End of Cheap Oil. *Scientific American.* March: 60–65.

Caso-Chávez, M., I. Pisanty, and E. Ezcurra (eds.). 2004. *Diagnóstico ambiental del Golfo de México.* Instituto Nacional de Ecología (INECOL A.C.) and Harte Research Institute for Gulf of Mexico Studies TAMU–CC, 2 vols. México City: Secretaría de Medio Ambiente y Recursos Naturales (SEMARNAT). 1108 pp.

CEC/NAAEC/NAFTA. 1997. *Ecological Regions of North America: Toward a Common Perspective.* Montreal: Commission for Environmental Cooperation, North American Agreement for Environmental Cooperation. 71 pp. and maps.

———. 2002. *Mapping Marine and Estuarine Ecological Regions of North America.* Commission for Environmental Cooperation, North American Agreement for Environmental Cooperation. Draft Workshop Report, Charleston, South Carolina, October 2002. Montreal.

Christie, P., and A. T. White. 1997. Trends in Development of Coastal Area Management in Tropical Countries: From Central to Community Orientation. *Coastal Management* 25 (2): 155–81.

———. 2000. Introduction to special issue on tropical coastal management. *Coastal Management* 28: 1–4.

Cicin-Sain, B., and R. W. Knecht. 1998. *Integrated Coastal and Ocean Management: Concepts and Practice.* Washington, D.C.: Island Press. 518 pp.

CONAPESCA. 2003. *Annual Fish Statistics.* http://www.conapesca.sagarpa.gob.mx/work/sites/cona/resources/LocalContent/2786/1/anuario2003.zip.

CONAPO, 2000. *Proyecciones de la población de México 2000–2025 estatales y nacional.* Consejo Nacional de Población. México City. http://www.conapo.gob.mx/m_en_cifras/principal.html.

Costanza, R., R. d'Arge, R. de Groot, S. Farber, M. Grasso, B. Hannon, S. Naeem, K. Limburg, J. Paruelo, R. V. O'Neill, R. Raskin, P. Sutton, and M. van den Belt. 1997. The Value of the World's Ecosystem Services and Natural Capital. *Nature* 387: 253.

Costanza, R., S. C. Farber, and J. Maxwell. 1989. The Valuation and Management of Wetland Ecosystems. *Ecological Economics* 1: 335–61.

Day, J. W., and A. Yáñez-Arancibia (eds.). 2008. *The Gulf of Mexico: Ecosystem-Based Management.* Harte Research Institute for Gulf of Mexico Studies. College Station: Texas A&M University Press.

Day, J. W., and P. H. Templet. 1989. Consequences of Sea-Level Rise: Implications from the Mississippi Delta. *Coastal Management* 17: 241–57.

Day, J. W., A. Díaz de León, G. González, P. Moreno, and A. Yáñez-Arancibia. 2004b. Diagnóstico ambiental del Golfo de México: Resumen ejecutivo. Pp. 15–44 in M. Caso-Chávez, I. Pisanty, and E. Ezcurra (eds.), *Diagnóstico ambiental del Golfo de México.* INE, INECOL A.C., Harte Research Institute for Gulf of Mexico Studies TAMU–CC, vol. 1. Mexico City: Secretaría de Medio Ambiente y Recursos Naturales (SEMARNAT).

Day, J. W., A. Yáñez-Arancibia, W. J. Mitsch, A. L. Lara, J. N. Day, J. Y. Ko, R. Lane, J. Lindsey, and D. Zárate. 2003. Using Ecotechnology to Address Water Quality and Wetland Habitat Loss Problems in the Mississippi Basin (and Grijalva/Usumacinta Basin): A Hierarchical Approach. *Biotechnology Advances* 22: 135–59.

Day, J. W., P. H. Templet, J. Y. Ko, W. J. Mitsch, G. P. Kemp, J. Johnston, G. Steyer, J. Barras, D. Justic, E. Clairain, and R. Theriot. 2004c. El delta del Mississippi: Funcionamiento del sistema, impactos ambientales, y desarrollo sustentable. Pp. 851–82 in M. Caso-Chávez, I. Pisanty, and E. Ezcurra (eds.), *Diagnóstico ambiental del Golfo de México.* INE, INECOL A.C., Harte Research Institute for Gulf of Mexico Studies TAMU–CC, vol. 2. Mexico City: Secretaría de Medio Ambiente y Recursos Naturales (SEMARNAT).

Day, J. W., J. Barras, E. Clairain, J. Johnston, D. Justic, P. Kemp, J. Y. Ko, R. R. Lane, W. J. Mitsch, G. Steyer, P. H. Templet, and A. Yáñez-Arancibia. 2005. Implications of Global Climatic Change and Energy Cost and Availability for the Restoration of the Mississippi Delta. *Ecological Engineering* 24 (4): 253–65.

Day, J. W., J. F. Martin, L. Cardoch, and P. H. Templet. 1997. System Functioning as a

Basis for Sustainable Management of Deltaic Ecosystems. *Coastal Management* 25 (2): 115–54.

Day, J. W., J. Y. Ko, J. Rybczyk, D. Sabins, R. Bean, G. Berthelot, C. Brantley, L. Cardoch, W. Conner, J. N. Day, A. J. Englande, S. Feagly, E. Hyfield, R. Lane, J. Lindsey, W. J. Mitsch, E. Reyes, and R. Twilley. 2004a. The Use of Wetlands in the Mississippi Delta for Wastewater Assimilation: A Review. *Ocean & Coastal Management* 47 (11–12): 671–92.

Deffeyes, K. S. 2001. *Hubbert's Peak: The Impending World Oil Shortage.* Princeton, N.J.: Princeton University Press. 208 pp.

———. 2002. World's Oil Production Peak Reckoned in Near Future. *Oil & Gas Journal* 100 (46): 46–48.

Delta Assembly. 1992. *Preserving the Global Environment: Developing a Shared Vision for the Gulf of Mexico.* Final Statement, College of Urban and Public Affairs, University of New Orleans. 36 pp.

Emanuel, K., 2005. Increasing Destructiveness of Tropical Cyclones over the Last 30 Years. *Nature* 436: 686.

Frugé, D. J .1999. United States Fish and Wildlife Service's Approach to Resource Management in the Gulf of Mexico River Drainages. Pp. 645–66 in H. Kumpf, K. Steidinger, and K. Sherman (eds.), *The Gulf of Mexico Large Marine Ecosystem: Assessment, Sustainability and Management.* Malden, Mass.: Blackwell Science.

Gold, G. 2004. Hidrocarburos en el sur del Golfo de México. Pp. 657–82 in M. Caso-Chávez, I. Pisanty, and E. Ezcurra (eds.), *Diagnóstico ambiental del Golfo de México.* INE, INECOL A.C., Harte Research Institute for Gulf of Mexico Studies TAMU–CC, vol. 1. Mexico City: Secretaría de Medio Ambiente y Recursos Naturales (SEMARNAT).

Hall, C.A.S., P. Tharakan, J. Hallock, C. Cleveland, and M. Jefferson. 2003. Hydrocarbons and the Evolution of Human Culture. *Nature* 426: 318–22.

Harwell, M. A., 1998. Science and Environmental Decision Making in South Florida. *Ecological Applications* 8 (3): 580–90.

Herrera-Silveira, J. A., N. A. Cirerol, L. Trocoli, F. A. Comin, and C. J. Madden. 2004. Eutrofización costera en la Península de Yucatán. Pp. 823–50 in M. Caso-Chávez, I. Pisanty, and E. Ezcurra (eds.), *Diagnóstico ambiental del Golfo de México.* INE, INECOL A.C., Harte Research Institute for Gulf of Mexico Studies TAMU–CC, vol. 2. Mexico City: Secretaría de Medio Ambiente y Recursos Naturales (SEMARNAT).

INEGI.1999. *Sistema de cuentas económicas y ecológicas de México 1993–1997.* Instituto Nacional de Estadística Geografía e Informática. Aguascalientes, México. URL: www.inegi.gob.mx.

———. 2001. *XII Censo general de población y vivienda 2000.* Instituto Nacional de Estadística Geografía e Informática. Aguascalientes, México. URL: www.inegi .gob.mx.

INEGI/INE-SEMARNAT. 2000. *Sustainable Development Indicators of Mexico.* Instituto Nacional de Estadística Geografía e Informática. Aguascalientes, México. URL: http://www.ine.gob.mx/ueajei/publicaciones/consultaPublicacion.html?id _pub=322&id_tema=12&dir=Consultas.

IPCC (Intergovernmental Panel on Climate Change). 2001. *Climate Change 2001: The Scientific Basis Contribution of Working Group 1 to the Third Assessment Report.* Cambridge, U.K.: Cambridge University Press.

Karr, J. R., and D. R. Dudley. 1981. Ecological Perspective on Water Quality Goals. *Environmental Management* 5 (1): 55–68.

Ko, J. Y., and J. W. Day. 2004. A Review of Ecological Impacts of Oil and Gas Development on Coastal Ecosystems in the Mississippi Delta. *Ocean & Coastal Management* 47 (11–12): 597–624.

Kumpf, H., K. Steidinger, and K. Sherman (eds.). 1999. *The Gulf of Mexico Large Marine Ecosystem: Assessment, Sustainability and Management.* Malden, Mass.: Blackwell Science. 704 pp.

León, C., and H. Rodríguez, 2004. Ambivalencias y asimetrías en el proceso de urbanización en el Golfo de México: Presión ambiental y concentración demográfica. Pp.1043–82 in M. Caso-Chávez, I. Pisanty, and E. Ezcurra (eds.), *Diagnóstico ambiental del Golfo de México.* INE, INECOL A.C., Harte Research Institute for Gulf of Mexico Studies TAMU–CC, vol. 2. Mexico City: Secretaría de Medio Ambiente y Recursos Naturales (SEMARNAT).

LeRoy Poff, N., M. M. Brinson, and J. W. Day. 2002. *Aquatic Ecosystems and Global Climate Change: Potential Impacts on Inland Freshwater and Coastal Wetlands Ecosystems in the United States.* Arlington, Va.: Pew Center on Global Climate Change. 44 pp.

Lohrenz, S. E., D. A. Weisenburg, R. A. Arnone, and X. Chen. 1999. What Controls Primary Production in the Gulf of Mexico? Pp. 151–70 in H. Kumpf, K. Steidinger, and K. Sherman (eds.), *The Gulf of Mexico Large Marine Ecosystem: Assessment, Sustainability and Management.* Malden, Mass.: Blackwell Science. 704 pp.

Maul, G. (ed.). 1993. *Climate Change in the Intra-Americas Sea.* London: Edward Arnold. 400 pp.

Mitsch, W. J., J. W. Day, J. Gilliam, P. Groffman, D. Hey, G. Randall, and N. Wong. 2001. Reducing Nitrogen Loading to the Gulf of Mexico from the Mississippi River Basin: Strategies to Counter a Persistent Problem. *BioScience* 51(5): 373–88.

Mitsch, W. J., and S.E. Jørgensen. 2003. *Ecological Engineering and Ecosystem Restoration.* New York: John Wiley and Sons. 411 pp.

Ning, Z. H., R. E. Turner, T. Doyle, and K. Abdollahi. 2003. *Integrated Assessment of the Climate Change Impacts on the Gulf Coast Region.* Baton Rouge, La.: United States Environmental Protection Agency and United States Geological Survey. 236 pp.

Odum, H. T. 1971. *Environment, Power and Society.* New York: John Wiley and Sons. 331 pp.

Olsen, S., and P. Christie. 2000. What Can We Learn from Tropical Coastal Management Experiences? *Coastal Management* 28 (1): 5–18.

Ortiz Pérez, M. A., A. P. Méndez-Linares, and J. R. Hernández-Santana. 2008. Sea-Level Rise and Vulnerability of Coastal Low-Land in the Mexican Area of the Gulf of Mexico and the Caribbean Sea. Chap. 15 in J. W. Day and A. Yáñez-Arancibia (eds.), *The Gulf of Mexico: Ecosystem-Based Management.* College Station: Texas A&M University Press.

PEMEX-PEP. 2004a. *Annual Report.* http://www.pep.pemex.com/reporteanual2004/webreporteingles/recursos/html/explotacion/prod_crudo_gas.htm.

———. 2004b. *Informe de Indicadores Petroleros.* http://www.pemex.com/index.cfm/action/contents/sectionID/1/catID/237/index.cfm?action=contents&sectionID=1&catID=237.

———. 2005. *Annual Report.* http://www.pep.pemex.com/reporteanual2005/webreporteingles/recursos/html/explotacion/prod_crudo_gas.htm.

Pickering, D. R., J. Kentor, and T. Ramadrishnan. 2003. *A Primer on Oil and Gas in Mexico.* Houston: Simmons and Company International. 32 pp.

PRC/INECOL. 2001. *Banco de datos Socioeconómicos y ambientales de la zona costera del Golfo de México y Mar Caribe.* Proyecto Sustentabilidad Ambiental del Desarrollo Económico de la Zona Costera del Golfo de México y Mar Caribe. Informe Técnico Instituto de Ecología A.C. Xalapa, México: CONACYT. 250 pp.

Rabalais, N. N., R. E. Turner, and W. J. Wiseman. 1999. Hypoxia in the Northern Gulf of Mexico: Linkage with the Mississippi River. Pp. 297–322 in H. Kumpf, H., K. Steidinger, and K. Sherman (eds.). *The Gulf of Mexico Large Marine Ecosystem: Assessment, Sustainability and Management.* Malden, Mass.: Blackwell Science.

Ramírez-Gordillo, J. 2003. El significado económico de la zona costera del estado de Veracruz. Tesis Profesional, Universidad Veracruzana, México. 93 pp.

Reuters. 2004. Mexico's PEMEX reports huge new oil finds. August 29.

Reyes, E., J. W. Day, A. L. Lara-Dominguez, P. Sánchez-Gil, D. Zárate, and A. Yáñez-Arancibia. 2004. Assessing Coastal Management Plans Using Watershed Spatial Model for the Mississippi Delta, USA, and the Usumacinta-Grijalva Delta, Mexico. *Ocean & Coastal Management* 47 (11–12): 693–708.

Reyes, E., J. W. Day, M. L. White, and A. Yáñez-Arancibia. 1993. Ecological and Resources Management Information Transfer for Laguna de Terminos, Mexico: A Computerized Interface. *Coastal Management* 21 (1): 37–52.

Rueda, M., 2005. Overdrive. *Latin Trade,* July.

Sánchez-Gil, P., and A. Yáñez-Arancibia. 1997. Grupos ecológicos funcionales y recursos pesqueros tropicales. Pp. 357–89 in D. Flores Hernández, P. Sánchez-Gil,

J. C. Seijo, and F. Arreguin (eds.), *Análisis y diagnóstico de los recursos pesqueros críticos del Golfo de México*. Universidad A. de Campeche EPOMEX Serie Científica 7. 496 pp.

Sánchez-Gil, P., A. Yáñez-Arancibia, J. Ramírez-Gordillo, J. W. Day, and P. H. Templet. 2004. Some Socio-Economic Indicators in the Mexican States of the Gulf of Mexico. *Ocean & Coastal Management* 47 (11–12): 581–97.

Scavia, D., J. C. Field, D. F. Boesch, R. W. Buddemeier, V. Burkett, D. R. Cayan, M. Fogarty, M. A. Harwell, R. W. Howarth, C. Mason, D. J. Reed, T. C. Royer, A. H. Sallenger, and J. G. Titus. 2002. Climate Change Impacts on U.S. Coastal and Marine Ecosystems. *Estuaries* 25 (2): 149–64.

SECTUR. 2006. *Secretaría de Turismo.* http://datatur.sectur.gob.mx/jsp/index.jsp.

Twilley, R. R., E. J. Barron, H. L. Gholz, M. A. Harwell, R. L. Miller, D. J. Reed, J. B. Rose, E. H. Siemann, R. G. Wetzel, and R. J. Zimmerman. 2001. *Confronting Climate Change in the Gulf Coast Region: Prospects for Sustaining Our Ecological Heritage.* Cambridge, Mass.: Union of Concerned Scientists and Ecological Society of America. 82 pp.

USFWS (U.S. Fish and Wildlife Service). 1995. *Watershed Based Ecosystems* (map). Washington, D.C.: United States Fish and Wildlife Service.

Westmacott, S. 2002. Where Should the Focus Be in Tropical Integrated Coastal Management? *Coastal Management* 30 (1): 67–84.

Windevoxhel, N. J., J. J. Rodríguez, and E. J. Lahmann. 1999. Situation of Integrated Coastal Management in Central America: Experiences of the IUCN Wetlands and Coastal Zone Conservation Program. *Ocean & Coastal Management* 42 (1–2): 257–82.

Woodley, S. 1993. Monitoring and Measuring Ecosystem Integrity in Canadian National Parks. Chap. 9 in S. Woodley, J. Kay and G. Francis (eds.), *Ecological Integrity and the Management of Ecosystems.* Ottawa: University of Waterloo and Canadian Parks Service.

Yáñez-Arancibia, A. 1999. Terms of Reference towards Coastal Management and Sustainable Development in Latin America: Introduction to Special Issue on Progress and Experiences. *Ocean & Coastal Management* 42 (91–92): 77–104.

Yáñez-Arancibia, A. 2000. Coastal Management in Latin America. Chap. 28, pp. 447–456 in C. Sheppard (ed.), *The Seas at the Millennium: An Environmental Evaluation,* 3 vols. Oxford, U.K.: Elsevier Science.

Yáñez-Arancibia, A., and J. W. Day. 2004. Environmental Sub-Regions in the Gulf of Mexico Coastal Zone: The Ecosystem Approach as an Integrated Management Tool. *Ocean & Coastal Management* 47 (11–12): 727–57.

———. 2005. Ecosistemas vulnerables, riesgo ecológico y el record 2005 de huracanes en el Golfo de México y Mar Caribe. http://www.ine.gob.mx/download/huracanes 2005.pdf.

Yáñez-Arancibia, A., A. L. Lara-Domínguez, P. Sánchez-Gil, and J. W. Day. 2004. Interacciones ecológicas estuario-mar: Marco conceptual para el manejo ambiental costero. Pp. 431–490 in M. Caso-Chávez, I. Pisanty, and E. Ezcurra (eds.), *Diagnóstico ambiental del Golfo de México.* INE, INECOL A.C., Harte Research Institute for Gulf of Mexico Studies TAMU–CC, vol. 1. Mexico City: Secretaría de Medio Ambiente y Recursos Naturales (SEMARNAT).

Yáñez-Arancibia, A., A. L. Lara, D. Zárate Lomelí, P. Sánchez-Gil, S. Jiménez, A. Sánchez, E. Rivera, A. Flores Nava, M. A. Ortiz Pérez, C. Muñoz, M. Becerra, J. W. Day, G. L. Powell, C. J. Madden, E. Reyes, and C. Barrientos (eds.). 2003. *Sub-region 2 Gulf of Mexico: Scaling, Scoping and Detailed Assessment.* Global International Water Assessment Report GIWA. Kalmar, Sweden: GEF-UNEP. 396 pp.

Yáñez-Arancibia, A., P. Bacon, M. Herzig, and E. Lahmann. 2006. Wetlands in Mexico, Central America and the Caribbean. Pp. 92–103 in P. Dugan (ed.), *Guide to Wetlands.* New York: Firefly Books. 304 pp.

Zárate Lomelí, D., and A. Yáñez-Arancibia (eds.). 2003. *Conclusiones: Necesidades para la gestión y el manejo integrado de la zona costera del Golfo de México y Mar Caribe,* SEMARNAT Technical Report. Mexico City: Louisiana State University, Baton Rouge, La., and INECOL A.C., Xalapa, Mexico. 16 pp.

Zárate Lomelí, D., and A. Yáñez-Arancibia. 2004. Plan Puebla-Panamá, los recursos naturales y el desarrollo sustentable: Planteamiento ecosistémico para la planificación ambiental. Unpublished.

Zárate Lomelí, D., T. Saavedra, J. L. Rojas, A. Yáñez-Arancibia, and E. Rivera. 1999. Terms of Reference towards an Integrated Management Policy in the Coastal Zone of the Gulf of Mexico and the Caribbean. *Ocean & Coastal Management* 42 (1–2): 345–68.

Zárate Lomelí, D., A. Yáñez-Arancibia, J. W. Day, M. Ortiz, A. L. Lara, C. Ojeda, L. J. Morales, and S. Guevara. 2004. Lineamientos para el programa regional de manejo integrado de la zona costera del Golfo de México y el Caribe. Pp. 899–936 in M. Caso-Chávez, I. Pisanty, and E. Ezcurra (eds.), *Diagnóstico ambiental del Golfo de México*. INE, INECOL A.C., Harte Research Institute for Gulf of Mexico Studies TAMU–CC, vol. 2. Mexico City: Secretaría de Medio Ambiente y Recursos Naturales (SEMARNAT).

Contributors

Charles M. Adams, Professor and Florida Sea Grant Extension Economist, Food and Resource Economics Department, University of Florida, P.O. Box 110240, Gainesville, Florida 32611-0240, cmadams@ufl.edu

James C. Cato, Senior Associate Dean and Director, School of Natural Resources and Environment, Director, Florida Sea Grant Program, Professor, Food and Resource Economics, University of Florida, P.O. Box 110400, Gainesville, Florida 32611-0400, jccato@ufl.edu

Charles S. Colgan, Professor of Public Policy and Management, Lead Economist, Market Data, National Ocean Economics Program, Edmund S. Muskie School of Public Service, University of Southern Maine, Portland, Maine 04104, colgan@usm.maine.edu

John W. Day Jr., Distinguished Professor Emeritus, Department of Oceanography and Coastal Sciences, School of the Coast and Environment, Louisiana State University, Baton Rouge, Louisiana 70803, johnday@lsu.edu

Emilio Hernandez, Ph.D. Student, Agricultural, Environmental and Development Economics, 241N Agricultural Administration, 2120 Fyffe Road, Columbus, Ohio 43202, Hernandez.162@osu.edu

Judith T. Kildow, Social Scientist and Director, National Ocean Economics Program, Monterey Bay Aquarium Research Institute, 7700 Sandholdt Road, Moss Landing, California 95039, jtk@mbari.org

Jim Lee, Professor, Department of Finance, Economics and Decision Sciences, College of Business, Texas A&M University–Corpus Christi, Corpus Christi, Texas 78412, jim.lee@tamucc.edu

Terry L. McCoy, Professor, Latin American Studies and Political Science, Director, Latin America Business Environment Program, Associate Director, Center for International Business Education and Research, University of Florida, P.O. Box 115530, Gainesville, Florida 32611-5330, tlmccoy@latam.ufl.edu

Linwood Pendleton, Associate Professor, Environmental Science and Engineering Program, Lead Non-Market Environmental Economist, National Ocean Economics Program, University of California–Los Angeles, Los Angeles, California 90095, linwoodp@ucla.edu

José J. Ramírez-Gordillo, Instituto de Ecología, A. C., Km 2.5 Carretera Antigua Coatepec 351, Congregación El Haya, CP 91070, Xalapa, Veracruz, Mexico, Jose.jesus@inecol.edu.mx

Alejandro Yáñez-Arancibia, Senior Scientist and Professor, Instituto de Ecología, A. C., Km 2.5 Carretera Antigua Coatepec 351, Congregación El Haya, CP 91070, Xalapa, Veracruz, Mexico, Alejandro.yanez@inecol.edu.mx

David W. Yoskowitz, Associate Professor of Economics, Harte Research Institute for Gulf of Mexico Studies and College of Business, Texas A&M University–Corpus Christi, Corpus Christi, Texas 78412, David.yoskowitz@tamucc.edu

Index